Praise for *Human C*

Chosen as a Best Book of 2019 by *Forbes*, *Financial Times*,
The Guardian, and *Daily Telegraph*

"This is the most important book I have read in quite some time. It lucidly
explains how the coming age of artificial superintelligence threatens human
control. Crucially, it also introduces a novel solution and a reason for hope."
—Daniel Kahneman, winner of the Nobel Prize
and author of *Thinking, Fast and Slow*

"A thought-provoking and highly readable account of the past, present and
future of AI. . . . Russell is grounded in the realities of the technology, including its many limitations, and isn't one to jump at the overheated language of sci-fi . . . If you are looking for a serious overview to the subject
that doesn't talk down to its non-technical readers, this is a good place to
start." —*Financial Times*

"Surely the most important book on AI this year. . . . A wry and witty tour
of intelligence and where it may take us." —*The Guardian*

"Worth reading *Human Compatible* by Stuart Russell (he's great) about
future AI risks and solutions." —Elon Musk

"A carefully written explanation of the concepts underlying AI as well as
the history of their development. If you want to understand how fast AI
is developing and why the technology is so dangerous, *Human Compatible*
is your guide." —*TechCrunch*

"I just finished Stuart Russell's marvelous book on AI safety, *Human Compatible*, and I can't recommend it highly enough!"
—Tim O'Reilly, founder and CEO of O'Reilly Media

"Before AI can solve the world's problems, it must overcome the challenge
of understanding humans—a feat we ourselves won't achieve soon . . .
Mr. Russell's exciting book goes deep, while sparkling with dry witticisms."
—*The Wall Street Journal*

"Stuart Russell has long been the most sensible voice in computer science on the topic of AI risk. And he has now written the book we've all been waiting for—a brilliant and utterly accessible guide to what will be either the best or worst technological development in human history."

—Sam Harris, author of *Waking Up* and host of the *Making Sense* podcast

"*Human Compatible* made me a convert to Russell's concerns with our ability to control our upcoming creation—superintelligent machines. Unlike outside alarmists and futurists, Russell is a leading authority on AI. His new book will educate the public about AI more than any book I can think of, and is a delightful and uplifting read."

—Judea Pearl, Turing Award winner and coauthor of *The Book of Why*

"A slew of books over recent years—notably Max Tegmark's *Life 3.0* and Nick Bostrom's *Superintelligence*—have warned of the dangers of runaway AI escaping our control . . . It is Russell, in his new book, *Human Compatible: Artificial Intelligence and the Problem of Control*, who has begun to calculate some solutions to this looming crisis, which we must solve, not just to survive, but also to harness the immense possibilities for good of AI . . . Russell's epiphany may be celebrated as the moment that humanity began the practical process of defending itself from the most powerful adversary it has ever faced."

—*The Telegraph*

"Stuart Russell wrote the book on AI and is leading the fight to change how we build it."

—*Vox*

"An AI expert's chilling warning . . . Fascinating, and significant . . . Russell is not warning of the dangers of conscious machines, just that superintelligent ones might be misused or might misuse themselves."

—*The Times*

"Stuart Russell, one of the most important AI scientists of the last twenty-five years, may have written the most important book about AI so far, on one of the most important questions of the twenty-first century: How to build AI to be compatible with us? The book proposes a novel and intriguing solution for this problem, while offering many thought-provoking ideas and insights about AI along the way. An accessible and engaging must-read for the developers of AI and the users of AI—that is, for all of us."

—James Manyika, chairman and director of the
McKinsey Global Institute

"A brilliantly clear and fascinating exposition of the history of computing thus far, and how very difficult true AI will be to build." —*The Spectator*

"A remarkable achievement and a must-read for anyone who is interested in this most important technology of our time. Russell is not someone with an ax to grind but someone who clearly feels a deep weight of responsibility as his own field gets closer to achieving its goals. Refreshingly, his message is not one of doom and gloom but of explaining the nature of the dangers and difficulties, before setting out a plan for how we can rise to these challenges. —Toby Ord, author of *The Precipice: Existential Risk and the Future of Humanity*

"Fascinating and significant. The AI industry is starting to accept that moving fast and breaking things is unwise if the thing that might break is humanity itself." —*The Sunday Times*

"This beautifully written book addresses a fundamental challenge for humanity: increasingly intelligent machines that do what we ask but not what we really intend. Essential reading if you care about our future."
—Yoshua Bengio, Turing Award winner and coauthor of *Deep Learning*

"Sound[s] an important alarm bell . . . *Human Compatible* marks a major stride in AI studies, not least in its emphasis on ethics. At the book's heart, Russell incisively discusses the misuses of AI." —*Nature*

"The same mix of demystifying authority and practical advice that Dr. Benjamin Spock once brought to the care and raising of children, Dr. Stuart Russell now brings to the care, raising, and, yes, disciplining of machines. He has written the book that most—but perhaps not all—machines would like you to read." —George Dyson, author of *Turing's Cathedral*

"Persuasively argued and lucidly imagined, *Human Compatible* offers an unflinching, incisive look at what awaits us in the decades ahead. No researcher has argued more persuasively about the risks of AI or shown more clearly the way forward. Anyone who takes the future seriously should pay attention."   —Brian Christian, author of *Algorithms to Live By*

"The right guide at the right time for technology enthusiasts seeking to explore the primary concepts of what makes AI valuable while simultaneously examining the disconcerting aspects of AI misuse."   —*Library Journal*

"A book that charts humanity's quest to understand intelligence, pinpoints why it became unsafe, and shows how to course-correct if we want to survive as a species. Stuart Russell, author of the leading AI textbook, can do all that with the wealth of knowledge of a prominent AI researcher and the persuasive clarity and wit of a brilliant educator."
                              —Jann Tallinn, cofounder of Skype

"Can we coexist happily with the intelligent machines that humans will create? 'Yes,' answers *Human Compatible*, 'but first . . .' Through a brilliant reimagining of the foundations of artificial intelligence, Russell takes you on a journey from the very beginning, explaining the questions raised by an AI-driven society, and beautifully making the case for how to ensure machines remain beneficial to humans. A totally readable and crucially important guide to the future from one of the world's leading experts."
                —Tabitha Goldstaub, cofounder of CognitionX and
                      head of the UK Government's AI Council

"In clear and compelling language, Stuart Russell describes the huge potential benefits of artificial intelligence, as well as the hazards and ethical challenges. It's especially welcome that a respected leading authority should offer this balanced appraisal, avoiding both hype and scaremongering."
                —Lord Martin Rees, Astronomer Royal and former
                      president of the Royal Society

PENGUIN BOOKS

HUMAN COMPATIBLE

Stuart Russell is a professor of computer science and holder of the Smith-Zadeh Chair in Engineering at the University of California, Berkeley. He has served as the vice-chair of the World Economic Forum's Council on AI and Robotics and as an adviser to the United Nations on arms control. He is an Andrew Carnegie fellow, as well as a fellow of the Association for the Advancement of Artificial Intelligence, the Association for Computing Machinery, and the American Association for the Advancement of Science. He is the author (with Peter Norvig) of the definitive and universally acclaimed textbook on AI, *Artificial Intelligence: A Modern Approach*.

ALSO BY STUART RUSSELL

The Use of Knowledge in Analogy and Induction (1989)

Do the Right Thing: Studies in Limited Rationality
(with Eric Wefald, 1991)

Artificial Intelligence: A Modern Approach
(with Peter Norvig, 1995, 2003, 2010, 2019)

Human Compatible

ARTIFICIAL INTELLIGENCE AND THE
PROBLEM OF CONTROL

Stuart Russell

PENGUIN BOOKS

PENGUIN BOOKS
An imprint of Penguin Random House LLC
penguinrandomhouse.com

First published in the United States of America by Viking,
an imprint of Penguin Random House LLC, 2019
Published in Penguin Books 2020

ISBN 9780525558637 (paperback)

LIBRARY OF CONGRESS CATALOGING-IN-PUBLICATION DATA
Names: Russell, Stuart J. (Stuart Jonathan), author.
Title: Human compatible : artificial intelligence and
the problem of control / Stuart Russell.
Description: [New York] : Viking, 2019. |
Includes bibliographical references and index.
Identifiers: LCCN 2019029688 (print) | LCCN 2019029689 (ebook) |
ISBN 9780525558613 (hardcover) | ISBN 9780525558620 (ebook)
Subjects: LCSH: Automation—Safety measures. |
Artificial intelligence—Social aspects. | Artificial intelligence—
Accidents—Risk assessment.
Classification: LCC Q334.7 .R87 2019 (print) |
LCC Q334.7 (ebook) | DDC 006.301—dc23
LC record available at https://lccn.loc.gov/2019029688
LC ebook record available at https://lccn.loc.gov/2019029689

Printed in the United States of America
6th Printing

Designed by Amanda Dewey

For Loy, Gordon, Lucy, George, and Isaac

CONTENTS

PREFACE

Why This Book? Why Now?

This book is about the past, present, and future of our attempt to understand and create intelligence. This matters, not because AI is rapidly becoming a pervasive aspect of the present but because it is the dominant technology of the future. The world's great powers are waking up to this fact, and the world's largest corporations have known it for some time. We cannot predict exactly how the technology will develop or on what timeline. Nevertheless, we must plan for the possibility that machines will far exceed the human capacity for decision making in the real world. What then?

Everything civilization has to offer is the product of our intelligence; gaining access to considerably greater intelligence would be the biggest event in human history. The purpose of the book is to explain why it might be the last event in human history and how to make sure that it is not.

Overview of the Book

The book has three parts. The first part (Chapters 1 to 3) explores the idea of intelligence in humans and in machines. The material requires no technical background, but for those who are interested, it is supplemented by four appendices that explain some of the core concepts underlying present-day AI systems. The second part (Chapters 4 to 6) discusses some problems arising from imbuing machines with intelligence. I focus in particular on the problem of control: retaining absolute power over machines that are more powerful than us. The third part (Chapters 7 to 10) suggests a new way to think about AI and to ensure that machines remain beneficial to humans, forever. The book is intended for a general audience but will, I hope, be of value in convincing specialists in artificial intelligence to rethink their fundamental assumptions.

Human
Compatible

1

IF WE SUCCEED

A long time ago, my parents lived in Birmingham, England, in a house near the university. They decided to move out of the city and sold the house to David Lodge, a professor of English literature. Lodge was by that time already a well-known novelist. I never met him, but I decided to read some of his books: *Changing Places* and *Small World*. Among the principal characters were fictional academics moving from a fictional version of Birmingham to a fictional version of Berkeley, California. As I was an actual academic from the actual Birmingham who had just moved to the actual Berkeley, it seemed that someone in the Department of Coincidences was telling me to pay attention.

One particular scene from *Small World* struck me: The protagonist, an aspiring literary theorist, attends a major international conference and asks a panel of leading figures, "What follows if everyone agrees with you?" The question causes consternation, because the panelists had been more concerned with intellectual combat than ascertaining truth or attaining understanding. It occurred to me then that an analogous question could be asked of the leading figures in AI: "What if you succeed?" The field's goal had always been to create human-level or superhuman AI, but there was little or no consideration of what would happen if we did.

A few years later, Peter Norvig and I began work on a new AI textbook, whose first edition appeared in 1995.[1] The book's final section is titled "What If We Do Succeed?" The section points to the possibility of good and bad outcomes but reaches no firm conclusions. By the time of the third edition in 2010, many people had finally begun to consider the possibility that superhuman AI might not be a good thing—but these people were mostly outsiders rather than mainstream AI researchers. By 2013, I became convinced that the issue not only belonged in the mainstream but was possibly the most important question facing humanity.

In November 2013, I gave a talk at the Dulwich Picture Gallery, a venerable art museum in south London. The audience consisted mostly of retired people—nonscientists with a general interest in intellectual matters—so I had to give a completely nontechnical talk. It seemed an appropriate venue to try out my ideas in public for the first time. After explaining what AI was about, I nominated five candidates for "biggest event in the future of humanity":

1. We all die (asteroid impact, climate catastrophe, pandemic, etc.).
2. We all live forever (medical solution to aging).
3. We invent faster-than-light travel and conquer the universe.
4. We are visited by a superior alien civilization.
5. We invent superintelligent AI.

I suggested that the fifth candidate, superintelligent AI, would be the winner, because it would help us avoid physical catastrophes and achieve eternal life and faster-than-light travel, if those were indeed possible. It would represent a huge leap—a discontinuity—in our civilization. The arrival of superintelligent AI is in many ways analogous to the arrival of a superior alien civilization but much more likely to occur. Perhaps most important, AI, unlike aliens, is something over which we have some say.

Then I asked the audience to imagine what would happen if we received notice from a superior alien civilization that they would arrive on Earth in thirty to fifty years. The word *pandemonium* doesn't begin to describe it. Yet our response to the anticipated arrival of superintelligent AI has been . . . well, underwhelming begins to describe it. (In a later talk, I illustrated this in the form of the email exchange shown in figure 1.) Finally, I explained the significance of superintelligent AI as follows: "Success would be the biggest event in human history . . . and perhaps the last event in human history."

From: Superior Alien Civilization <sac12@sirius.canismajor.u>

To: humanity@UN.org

Subject: Contact

Be warned: we shall arrive in 30–50 years

From: humanity@UN.org

To: Superior Alien Civilization <sac12@sirius.canismajor.u>

Subject: Out of office: Re: Contact

Humanity is currently out of the office. We will respond to your message when we return. ☺

FIGURE 1: Probably not the email exchange that would follow the first contact by a superior alien civilization.

A few months later, in April 2014, I was at a conference in Iceland and got a call from National Public Radio asking if they could interview me about the movie *Transcendence*, which had just been released in the United States. Although I had read the plot summaries and reviews, I hadn't seen it because I was living in Paris at the time, and it would not be released there until June. It so happened, however, that

I had just added a detour to Boston on the way home from Iceland, so that I could participate in a Defense Department meeting. So, after arriving at Boston's Logan Airport, I took a taxi to the nearest theater showing the movie. I sat in the second row and watched as a Berkeley AI professor, played by Johnny Depp, was gunned down by anti-AI activists worried about, yes, superintelligent AI. Involuntarily, I shrank down in my seat. (Another call from the Department of Coincidences?) Before Johnny Depp's character dies, his mind is uploaded to a quantum supercomputer and quickly outruns human capabilities, threatening to take over the world.

On April 19, 2014, a review of *Transcendence*, co-authored with physicists Max Tegmark, Frank Wilczek, and Stephen Hawking, appeared in the *Huffington Post*. It included the sentence from my Dulwich talk about the biggest event in human history. From then on, I would be publicly committed to the view that my own field of research posed a potential risk to my own species.

How Did We Get Here?

The roots of AI stretch far back into antiquity, but its "official" beginning was in 1956. Two young mathematicians, John McCarthy and Marvin Minsky, had persuaded Claude Shannon, already famous as the inventor of information theory, and Nathaniel Rochester, the designer of IBM's first commercial computer, to join them in organizing a summer program at Dartmouth College. The goal was stated as follows:

> The study is to proceed on the basis of the conjecture that every aspect of learning or any other feature of intelligence can in principle be so precisely described that a machine can be made to simulate it. An attempt will be made to find how to make machines use language, form abstractions and concepts, solve kinds of problems now reserved for humans, and improve themselves. We think

that a significant advance can be made in one or more of these problems if a carefully selected group of scientists work on it together for a summer.

Needless to say, it took much longer than a summer: we are still working on all these problems.

In the first decade or so after the Dartmouth meeting, AI had several major successes, including Alan Robinson's algorithm for general-purpose logical reasoning[2] and Arthur Samuel's checker-playing program, which taught itself to beat its creator.[3] The first AI bubble burst in the late 1960s, when early efforts at machine learning and machine translation failed to live up to expectations. A report commissioned by the UK government in 1973 concluded, "In no part of the field have the discoveries made so far produced the major impact that was then promised."[4] In other words, the machines just weren't smart enough.

My eleven-year-old self was, fortunately, unaware of this report. Two years later, when I was given a Sinclair Cambridge Programmable calculator, I just wanted to make it intelligent. With a maximum program size of thirty-six keystrokes, however, the Sinclair was not quite big enough for human-level AI. Undeterred, I gained access to the giant CDC 6600 supercomputer[5] at Imperial College London and wrote a chess program—a stack of punched cards two feet high. It wasn't very good, but it didn't matter. I knew what I wanted to do.

By the mid-1980s, I had become a professor at Berkeley, and AI was experiencing a huge revival thanks to the commercial potential of so-called expert systems. The second AI bubble burst when these systems proved to be inadequate for many of the tasks to which they were applied. Again, the machines just weren't smart enough. An AI winter ensued. My own AI course at Berkeley, currently bursting with over nine hundred students, had just twenty-five students in 1990.

The AI community learned its lesson: smarter, obviously, was better, but we would have to do our homework to make that happen. The field became far more mathematical. Connections were made to the long-established disciplines of probability, statistics, and control theory. The seeds of today's progress were sown during that AI winter, including early work on large-scale probabilistic reasoning systems and what later became known as *deep learning*.

Beginning around 2011, deep learning techniques began to produce dramatic advances in speech recognition, visual object recognition, and machine translation—three of the most important open problems in the field. By some measures, machines now match or exceed human capabilities in these areas. In 2016 and 2017, DeepMind's AlphaGo defeated Lee Sedol, former world Go champion, and Ke Jie, the current champion—events that some experts predicted wouldn't happen until 2097, if ever.[6]

Now AI generates front-page media coverage almost every day. Thousands of start-up companies have been created, fueled by a flood of venture funding. Millions of students have taken online AI and machine learning courses, and experts in the area command salaries in the millions of dollars. Investments flowing from venture funds, national governments, and major corporations are in the tens of billions of dollars annually—more money in the last five years than in the entire previous history of the field. Advances that are already in the pipeline, such as self-driving cars and intelligent personal assistants, are likely to have a substantial impact on the world over the next decade or so. The potential economic and social benefits of AI are vast, creating enormous momentum in the AI research enterprise.

What Happens Next?

Does this rapid rate of progress mean that we are about to be overtaken by machines? No. There are several breakthroughs that have to happen before we have anything resembling machines with superhuman intelligence.

Scientific breakthroughs are notoriously hard to predict. To get a sense of just how hard, we can look back at the history of another field with civilization-ending potential: nuclear physics.

In the early years of the twentieth century, perhaps no nuclear physicist was more distinguished than Ernest Rutherford, the discoverer of the proton and the "man who split the atom" (figure 2[a]). Like his colleagues, Rutherford had long been aware that atomic nuclei stored immense amounts of energy; yet the prevailing view was that tapping this source of energy was impossible.

On September 11, 1933, the British Association for the Advancement of Science held its annual meeting in Leicester. Lord Rutherford addressed the evening session. As he had done several times before, he poured cold water on the prospects for atomic energy: "Anyone who looks for a source of power in the transformation of the atoms is talking moonshine." Rutherford's speech was reported in the *Times* of London the next morning (figure 2[b]).

TRANSFORMATION OF ELEMENTS

FROM OUR SPECIAL CORRESPONDENTS

LEICESTER, SEPT. 11

What, Lord Rutherford asked in conclusion, were the prospects 20 or 30 years ahead ?

It was a very poor and inefficient way of producing energy, and anyone who looked for a source of power in the transformation of the atoms was talking moonshine.

(a)   (b)   (c)

FIGURE 2: (a) Lord Rutherford, nuclear physicist. (b) Excerpts from a report in the *Times* of September 12, 1933, concerning a speech given by Rutherford the previous evening. (c) Leo Szilard, nuclear physicist.

Leo Szilard (figure 2[c]), a Hungarian physicist who had recently fled from Nazi Germany, was staying at the Imperial Hotel on Russell

Square in London. He read the *Times*' report at breakfast. Mulling over what he had read, he went for a walk and invented the neutron-induced nuclear chain reaction.[7] The problem of liberating nuclear energy went from impossible to essentially solved in less than twenty-four hours. Szilard filed a secret patent for a nuclear reactor the following year. The first patent for a nuclear weapon was issued in France in 1939.

The moral of this story is that betting against human ingenuity is foolhardy, particularly when our future is at stake. Within the AI community, a kind of denialism is emerging, even going as far as denying the possibility of success in achieving the long-term goals of AI. It's as if a bus driver, with all of humanity as passengers, said, "Yes, I am driving as hard as I can towards a cliff, but trust me, we'll run out of gas before we get there!"

I am not saying that success in AI will *necessarily* happen, and I think it's quite unlikely that it will happen in the next few years. It seems prudent, nonetheless, to prepare for the eventuality. If all goes well, it would herald a golden age for humanity, but we have to face the fact that we are planning to make entities that are far more powerful than humans. How do we ensure that they never, ever have power over us?

To get just an inkling of the fire we're playing with, consider how content-selection algorithms function on social media. They aren't particularly intelligent, but they are in a position to affect the entire world because they directly influence billions of people. Typically, such algorithms are designed to maximize *click-through*, that is, the probability that the user clicks on presented items. The solution is simply to present items that the user likes to click on, right? Wrong. The solution is to change the user's preferences so that they become more predictable. A more predictable user can be fed items that they are likely to click on, thereby generating more revenue. People with more extreme political views tend to be more predictable in which items they will click on. (Possibly there is a category of articles that die-hard centrists are likely to click on, but it's not easy to imagine what this category consists of.) Like any rational entity, the algorithm learns how to modify the state of its environment—in this case, the user's mind—in order to maximize its own reward.[8] The consequences include the resurgence of fascism, the dissolution of the social contract that underpins democracies around the world, and potentially the end of the European Union and NATO. Not bad for a few lines of code, even if it had a helping hand from some humans. Now imagine what a *really* intelligent algorithm would be able to do.

What Went Wrong?

The history of AI has been driven by a single mantra: "The more intelligent the better." I am convinced that this is a mistake—not because of some vague fear of being superseded but because of the way we have understood intelligence itself.

The concept of intelligence is central to who we are—that's why we call ourselves *Homo sapiens*, or "wise man." After more than two thousand years of self-examination, we have arrived at a characterization of intelligence that can be boiled down to this:

> *Humans are intelligent to the extent that our actions can be expected to achieve our objectives.*

All those other characteristics of intelligence—perceiving, thinking, learning, inventing, and so on—can be understood through their contributions to our ability to act successfully. From the very beginnings of AI, intelligence in machines has been defined in the same way:

> *Machines are intelligent to the extent that their actions can be expected to achieve their objectives.*

Because machines, unlike humans, have no objectives of their own, we give them objectives to achieve. In other words, we build optimizing machines, we feed objectives into them, and off they go.

This general approach is not unique to AI. It recurs throughout the technological and mathematical underpinnings of our society. In the field of control theory, which designs control systems for everything from jumbo jets to insulin pumps, the job of the system is to minimize a *cost function* that typically measures some deviation from a desired behavior. In the field of economics, mechanisms and policies are designed to maximize the *utility* of individuals, the *welfare* of groups, and the *profit* of corporations.[9] In operations research, which solves complex logistical and manufacturing problems, a solution maximizes an expected *sum of rewards* over time. Finally, in statistics, learning algorithms are designed to minimize an expected *loss function* that defines the cost of making prediction errors.

Evidently, this general scheme—which I will call the *standard model*—is widespread and extremely powerful. Unfortunately, *we don't want machines that are intelligent in this sense.*

The drawback of the standard model was pointed out in 1960 by Norbert Wiener, a legendary professor at MIT and one of the leading mathematicians of the mid-twentieth century. Wiener had just seen Arthur Samuel's checker-playing program learn to play checkers far better than its creator. That experience led him to write a prescient but little-known paper, "Some Moral and Technical Consequences of Automation."[10] Here's how he states the main point:

> If we use, to achieve our purposes, a mechanical agency with whose operation we cannot interfere effectively . . . we had better be quite sure that the purpose put into the machine is the purpose which we really desire.

"The purpose put into the machine" is exactly the objective that machines are optimizing in the standard model. If we put the wrong objective into a machine that is more intelligent than us, it will achieve the objective, and we lose. The social-media meltdown I described earlier is just a foretaste of this, resulting from optimizing the wrong objective on a global scale with fairly unintelligent algorithms. In Chapter 5, I spell out some far worse outcomes.

All this should come as no great surprise. For thousands of years, we have known the perils of getting exactly what you wish for. In every story where someone is granted three wishes, the third wish is always to undo the first two wishes.

In summary, it seems that the march towards superhuman intelligence is unstoppable, but success might be the undoing of the human race. Not all is lost, however. We have to understand where we went wrong and then fix it.

Can We Fix It?

The problem is right there in the basic definition of AI. We say that machines are intelligent to the extent that their actions can be expected to achieve *their* objectives, but we have no reliable way to make sure that *their* objectives are the same as *our* objectives.

What if, instead of allowing machines to pursue *their* objectives, we insist that they pursue *our* objectives? Such a machine, if it could be designed, would be not just *intelligent* but also *beneficial* to humans. So let's try this:

> Machines are **beneficial** to the extent that **their** actions can be expected to achieve **our** objectives.

This is probably what we should have done all along.

The difficult part, of course, is that our objectives are in us (all eight billion of us, in all our glorious variety) and not in the machines. It is, nonetheless, possible to build machines that are beneficial in exactly this sense. Inevitably, these machines will be uncertain about our objectives—after all, we are uncertain about them ourselves—but it turns out that this is a feature, not a bug (that is, a good thing and not a bad thing). Uncertainty about objectives implies that machines will necessarily defer to humans: they will ask permission, they will accept correction, and they will allow themselves to be switched off.

Removing the assumption that machines should have a definite objective means that we will need to tear out and replace part of the foundations of artificial intelligence—the basic definitions of what we are trying to do. That also means rebuilding a great deal of the superstructure—the accumulation of ideas and methods for actually doing AI. The result will be a new relationship between humans and machines, one that I hope will enable us to navigate the next few decades successfully.

2

INTELLIGENCE IN HUMANS AND MACHINES

When you arrive at a dead end, it's a good idea to retrace your steps and work out where you took a wrong turn. I have argued that the standard model of AI, wherein machines optimize a fixed objective supplied by humans, is a dead end. The problem is not that we might *fail* to do a good job of building AI systems; it's that we might *succeed* too well. The very definition of success in AI is wrong.

So let's retrace our steps, all the way to the beginning. Let's try to understand how our concept of intelligence came about and how it came to be applied to machines. Then we have a chance of coming up with a better definition of what counts as a good AI system.

Intelligence

How does the universe work? How did life begin? Where are my keys? These are fundamental questions worthy of thought. But who is asking these questions? How am I answering them? How can a handful of matter—the few pounds of pinkish-gray blancmange we call a brain—perceive, understand, predict, and manipulate a world of unimaginable vastness? Before long, the mind turns to examine itself.

We have been trying for thousands of years to understand how our minds work. Initially, the purposes included curiosity, self-management, persuasion, and the rather pragmatic goal of analyzing mathematical arguments. Yet every step towards an explanation of how the mind works is also a step towards the creation of the mind's capabilities in an artifact—that is, a step towards artificial intelligence.

Before we can understand how to create intelligence, it helps to understand what it is. The answer is not to be found in IQ tests, or even in Turing tests, but in a simple relationship between what we perceive, what we want, and what we do. Roughly speaking, an entity is intelligent to the extent that what it does is likely to achieve what it wants, given what it has perceived.

Evolutionary origins

Consider a lowly bacterium, such as *E. coli*. It is equipped with about half a dozen flagella—long, hairlike tentacles that rotate at the base either clockwise or counterclockwise. (The rotary motor itself is an amazing thing, but that's another story.) As *E. coli* floats about in its liquid home—your lower intestine—it alternates between rotating its flagella clockwise, causing it to "tumble" in place, and counterclockwise, causing the flagella to twine together into a kind of propeller so the bacterium swims in a straight line. Thus, *E. coli* does a sort of random walk—swim, tumble, swim, tumble—that allows it to find and consume glucose rather than staying put and dying of starvation.

If this were the whole story, we wouldn't say that *E. coli* is particularly intelligent, because its actions would not depend in any way on its environment. It wouldn't be making any decisions, just executing a fixed behavior that evolution has built into its genes. But this isn't the whole story. When *E. coli* senses an increasing concentration of glucose, it swims longer and tumbles less, and it does the opposite when it senses a decreasing concentration of glucose. So, what it does (swim towards glucose) is likely to achieve what it wants (more glucose, let's assume), given what it has perceived (an increasing glucose concentration).

Perhaps you are thinking, "But evolution built this into its genes too! How does that make it intelligent?" This is a dangerous line of reasoning, because evolution built the basic design of your brain into your genes too, and presumably you wouldn't wish to deny your own intelligence on that basis. The point is that what evolution has built into *E. coli*'s genes, as it has into yours, is a mechanism whereby the bacterium's behavior varies according to what it perceives in its environment. Evolution doesn't know, in advance, where the glucose is going to be or where your keys are, so putting the capability to find them into the organism is the next best thing.

Now, *E. coli* is no intellectual giant. As far as we know, it doesn't remember where it has been, so if it goes from A to B and finds no glucose, it's just as likely to go back to A. If we construct an environment where every attractive glucose gradient leads only to a spot of phenol (which is a poison for *E. coli*), the bacterium will keep following those gradients. It never learns. It has no brain, just a few simple chemical reactions to do the job.

A big step forward occurred with *action potentials*, which are a form of electrical signaling that first evolved in single-celled organisms around a billion years ago. Later multicellular organisms evolved specialized cells called *neurons* that use electrical action potentials to carry signals rapidly—up to 120 meters per second, or 270 miles per hour—within the organism. The connections between neurons are called *synapses*. The strength of the synaptic connection dictates how much electrical excitation passes from one neuron to another. By changing the strength of synaptic connections, animals learn.[1] Learning confers a huge evolutionary advantage, because the animal can adapt to a range of circumstances. Learning also speeds up the rate of evolution itself.

Initially, neurons were organized into *nerve nets*, which are distributed throughout the organism and serve to coordinate activities such as eating and digestion or the timed contraction of muscle cells across a wide area. The graceful propulsion of jellyfish is the result of a nerve net. Jellyfish have no brains at all.

Brains came later, along with complex sense organs such as eyes and ears. Several hundred million years after jellyfish emerged with their nerve nets, we humans arrived with our big brains—a hundred billion (10^{11}) neurons and a quadrillion (10^{15}) synapses. While slow compared to electronic circuits, the "cycle time" of a few milliseconds per state change is fast compared to most biological processes. The human brain is often described by its owners as "the most complex object in the universe," which probably isn't true but is a good excuse for the fact that we still understand little about how it really works. While we know a great deal about the biochemistry of neurons and synapses and the anatomical structures of the brain, the neural implementation of the *cognitive* level—learning, knowing, remembering, reasoning, planning, deciding, and so on—is still mostly anyone's guess.[2] (Perhaps that will change as we understand more about AI, or as we develop ever more precise tools for measuring brain activity.) So, when one reads in the media that such-and-such AI technique "works just like the human brain," one may suspect it's either just someone's guess or plain fiction.

In the area of *consciousness*, we really do know nothing, so I'm going to say nothing. No one in AI is working on making machines conscious, nor would anyone know where to start, and no behavior has consciousness as a prerequisite. Suppose I give you a program and ask, "Does this present a threat to humanity?" You analyze the code and indeed, when run, the code will form and carry out a plan whose result will be the destruction of the human race, just as a chess program will form and carry out a plan whose result will be the defeat of any human who faces it. Now suppose I tell you that the code, when run, also creates a form of machine consciousness. Will that change your prediction? Not at all. It makes *absolutely no difference.*[3] Your prediction about its behavior is exactly the same, because the prediction is based on the code. All those Hollywood plots about machines mysteriously becoming conscious and hating humans are really missing the point: it's competence, not consciousness, that matters.

There is one important cognitive aspect of the brain that we *are* beginning to understand—namely, the *reward system.* This is an internal signaling system, mediated by dopamine, that connects positive and negative stimuli to behavior. Its workings were discovered by the Swedish neuroscientist Nils-Åke Hillarp and his collaborators in the late 1950s. It causes us to seek out positive stimuli, such as sweet-tasting foods, that increase dopamine levels; it makes us avoid negative stimuli, such as hunger and pain, that decrease dopamine levels. In a sense it's quite similar to *E. coli*'s glucose-seeking mechanism, but much more complex. It comes with built-in methods for learning, so that our behavior becomes more effective at obtaining reward over time. It also allows for delayed gratification, so that we learn to desire things such as money that provide eventual reward rather than immediate reward. One reason we understand the brain's reward system is that it resembles the method of *reinforcement learning* developed in AI, for which we have a very solid theory.[4]

From an evolutionary point of view, we can think of the brain's reward system, just like *E. coli*'s glucose-seeking mechanism, as a way of improving evolutionary fitness. Organisms that are more effective in seeking reward—that is, finding delicious food, avoiding pain, engaging in sexual activity, and so on—are more likely to propagate their genes. It is extraordinarily difficult for an organism to decide what actions are most likely, in the long run, to result in successful propagation of its genes, so evolution has made it easier for us by providing built-in signposts.

These signposts are not perfect, however. There are ways to obtain reward that probably *reduce* the likelihood that one's genes will propagate. For example, taking drugs, drinking vast quantities of sugary carbonated beverages, and playing video games for eighteen hours a day all seem counterproductive in the reproduction stakes. Moreover, if you were given direct electrical access to your reward system, you would probably self-stimulate without stopping until you died.[5]

The misalignment of reward signals and evolutionary fitness doesn't affect only isolated individuals. On a small island off the coast of Panama lives the pygmy three-toed sloth, which appears to be addicted to a Valium-like substance in its diet of red mangrove leaves and may be going extinct.[6] Thus, it seems that an entire species can disappear if it finds an ecological niche where it can satisfy its reward system in a maladaptive way.

Barring these kinds of accidental failures, however, learning to maximize reward in natural environments will usually improve one's chances for propagating one's genes and for surviving environmental changes.

Evolutionary accelerator

Learning is good for more than surviving and prospering. It also *speeds up evolution*. How could this be? After all, learning doesn't change one's DNA, and evolution is all about changing DNA over generations. The connection between learning and evolution was proposed in 1896 by the American psychologist James Baldwin[7] and independently by the British ethologist Conwy Lloyd Morgan[8] but not generally accepted at the time.

The Baldwin effect, as it is now known, can be understood by imagining that evolution has a choice between creating an *instinctive* organism whose every response is fixed in advance and creating an *adaptive* organism that learns what actions to take. Now suppose, for the purposes of illustration, that the optimal instinctive organism can be coded as a six-digit number, say, 472116, while in the case of the adaptive organism, evolution specifies only 472*** and the organism itself has to fill in the last three digits by learning during its lifetime.

Clearly, if evolution has to worry about choosing only the first three digits, its job is much easier; the adaptive organism, in learning the last three digits, is doing in one lifetime what evolution would have taken many generations to do. So, provided the adaptive organisms can survive while learning, it seems that the capability for learning constitutes an evolutionary shortcut. Computational simulations suggest that the Baldwin effect is real.[9] The effects of culture only accelerate the process, because an organized civilization protects the individual organism while it is learning and passes on information that the individual would otherwise need to learn for itself.

The story of the Baldwin effect is fascinating but incomplete: it assumes that learning and evolution necessarily point in the same direction. That is, it assumes that whatever internal feedback signal defines the direction of learning within the organism is perfectly aligned with evolutionary fitness. As we have seen in the case of the pygmy three-toed sloth, this does not seem to be true. At best, built-in mechanisms for learning provide only a crude hint of the long-term consequences of any given action for evolutionary fitness. Moreover, one has to ask, "How did the reward system get there in the first place?" The answer, of course, is by an evolutionary process, one that internalized a feedback mechanism that is at least somewhat aligned with evolutionary fitness.[10] Clearly, a learning mechanism that caused organisms to run away from potential mates and towards predators would not last long.

Thus, we have the Baldwin effect to thank for the fact that neurons, with their capabilities for learning and problem solving, are so widespread in the animal kingdom. At the same time, it is important to understand that evolution doesn't really care whether you have a brain or think interesting thoughts. Evolution considers you only as an *agent*, that is, something that acts. Such worthy intellectual characteristics as logical reasoning, purposeful planning, wisdom, wit, imagination, and creativity may be essential for making an agent intelligent, or they may not. One reason artificial intelligence is so fascinating is that it offers a potential route to understanding these issues: we may come to understand both how these intellectual characteristics make intelligent behavior possible and why it's impossible to produce truly intelligent behavior without them.

Rationality for one

From the earliest beginnings of ancient Greek philosophy, the concept of intelligence has been tied to the ability to perceive, to reason, and to act *successfully*.[11] Over the centuries, the concept has become both broader in its applicability and more precise in its definition.

Aristotle, among others, studied the notion of successful reasoning—methods of logical deduction that would lead to true conclusions given true premises. He also studied the process of deciding how to act—sometimes called *practical reasoning*—and proposed that it involved deducing that a certain course of action would achieve a desired goal:

> We deliberate not about ends, but about means. For a doctor does not deliberate whether he shall heal, nor an orator whether he shall persuade. . . . They assume the end and consider how and by what means it is attained, and if it seems easily and best produced thereby; while if it is achieved by one means only they consider *how* it will be achieved by this and by what means *this* will be achieved, till they come to the first cause . . . and what is last in the order of analysis seems to be first in the order of becoming. And if we come on an impossibility, we give up the search, e.g., if we need money and this cannot be got; but if a thing appears possible we try to do it.[12]

This passage, one might argue, set the tone for the next two-thousand-odd years of Western thought about rationality. It says that the "end"—what the person wants—is fixed and given; and it says that the rational action is one that, according to logical deduction across a sequence of actions, "easily and best" produces the end.

Aristotle's proposal seems reasonable, but it isn't a complete guide to rational behavior. In particular, it omits the issue of uncertainty. In the real world, reality has a tendency to intervene, and few actions or sequences of actions are truly guaranteed to achieve the intended end. For example, it is a rainy Sunday in Paris as I write this sentence, and on Tuesday at 2:15 p.m. my flight to Rome leaves from Charles de Gaulle Airport, about forty-five minutes from my house. I plan to leave for the airport around 11:30 a.m., which should give me plenty of time, but it probably means at least an hour sitting in the departure area. Am I *certain* to catch the flight? Not at all. There could be huge traffic jams, the taxi drivers may be on strike, the taxi I'm in may break down or the driver may be arrested after a high-speed chase, and so on. Instead, I could leave for the airport on Monday, a whole day in advance. This would greatly reduce the chance of missing the flight, but the prospect of a night in the departure lounge is not an appealing one. In other words, my plan involves a *trade-off* between the certainty of success and the cost of ensuring that degree of certainty. The following plan for buying a house involves a similar trade-off: buy a lottery ticket, win a million dollars, then buy the house. This plan "easily and best" produces the end, but it's not very likely to succeed. The difference between this harebrained house-buying plan and my sober and sensible airport plan is, however, just a matter of degree. Both are gambles, but one seems more rational than the other.

It turns out that gambling played a central role in generalizing Aristotle's proposal to account for uncertainty. In the 1560s, the Italian mathematician Gerolamo Cardano developed the first mathematically precise theory of probability—using dice games as his main example. (Unfortunately, his work was not published until 1663.[13]) In the seventeenth century, French thinkers including Antoine Arnauld and Blaise Pascal began—for assuredly mathematical reasons—to

study the question of rational decisions in gambling.[14] Consider the
following two bets:

A: 20 percent chance of winning $10
B: 5 percent chance of winning $100

The proposal the mathematicians came up with is probably the same
one you would come up with: compare the *expected values* of the bets,
which means the average amount you would expect to get from each
bet. For bet A, the expected value is 20 percent of $10, or $2. For bet
B, the expected value is 5 percent of $100, or $5. So bet B is better,
according to this theory. The theory makes sense, because if the same
bets are offered over and over again, a bettor who follows the rule ends
up with more money than one who doesn't.

In the eighteenth century, the Swiss mathematician Daniel Bernoulli noticed that this rule didn't seem to work well for larger amounts
of money.[15] For example, consider the following two bets:

A: 100 percent chance of getting $10,000,000
(expected value $10,000,000)
B: 1 percent chance of getting $1,000,000,100
(expected value $10,000,001)

Most readers of this book, as well as its author, would prefer bet A to
bet B, even though the expected-value rule says the opposite! Bernoulli posited that bets are evaluated not according to expected monetary value but according to expected *utility*. Utility—the property of
being useful or beneficial to a person—was, he suggested, an internal,
subjective quantity related to, but distinct from, monetary value. In
particular, utility exhibits *diminishing returns with respect to money*.
This means that the utility of a given amount of money is not strictly
proportional to the amount but grows more slowly. For example, the
utility of having $1,000,000,100 is much less than a hundred times

the utility of having $10,000,000. How much less? You can ask yourself! What would the odds of winning a billion dollars have to be for you to give up a guaranteed ten million? I asked this question of the graduate students in my class and their answer was around 50 percent, meaning that bet B would have an expected value of $500 million to match the desirability of bet A. Let me say that again: bet B would have an expected dollar value fifty times greater than bet A, but the two bets would have equal utility.

Bernoulli's introduction of utility—an invisible property—to explain human behavior via a mathematical theory was an utterly remarkable proposal for its time. It was all the more remarkable for the fact that, unlike monetary amounts, the utility values of various bets and prizes are not directly observable; instead, utilities are to be *inferred* from the *preferences* exhibited by an individual. It would be two centuries before the implications of the idea were fully worked out and it became broadly accepted by statisticians and economists.

In the middle of the twentieth century, John von Neumann (a great mathematician after whom the standard "von Neumann architecture" for computers was named[16]) and Oskar Morgenstern published an *axiomatic* basis for utility theory.[17] What this means is the following: as long as the preferences exhibited by an individual satisfy certain basic axioms that any rational agent should satisfy, then *necessarily* the choices made by that individual can be described as maximizing the expected value of a utility function. In short, *a rational agent acts so as to maximize expected utility*.

It's hard to overstate the importance of this conclusion. In many ways, artificial intelligence has been mainly about working out the details of how to build rational machines.

Let's look in a bit more detail at the axioms that rational entities are expected to satisfy. Here's one, called *transitivity*: if you prefer A to B and you prefer B to C, then you prefer A to C. This seems pretty reasonable! (If you prefer sausage pizza to plain pizza, and you prefer plain pizza to pineapple pizza, then it seems reasonable to predict that you will choose sausage pizza over pineapple pizza.) Here's another, called *monotonicity*: if you prefer prize A to prize B, and you have a choice of lotteries where A and B are the only two possible outcomes, you prefer the lottery with the highest probability of getting A rather than B. Again, pretty reasonable.

Preferences are not just about pizza and lotteries with monetary prizes. They can be about anything at all; in particular, they can be about entire future lives and the lives of others. When dealing with preferences involving sequences of events over time, there is an additional assumption that is often made, called *stationarity*: if two different futures A and B begin with the same event, and you prefer A to B, you still prefer A to B after the event has occurred. This sounds reasonable, but it has a surprisingly strong consequence: the utility of any sequence of events is the sum of rewards associated with each event (possibly discounted over time, by a sort of mental interest rate).[18] Although this "utility as a sum of rewards" assumption is widespread—going back at least to the eighteenth-century "hedonic calculus" of Jeremy Bentham, the founder of utilitarianism—the stationarity assumption on which it is based is not a necessary property of rational agents. Stationarity also rules out the possibility that one's preferences might change over time, whereas our experience indicates otherwise.

Despite the reasonableness of the axioms and the importance of the conclusions that follow from them, utility theory has been subjected to a continual barrage of objections since it first became widely known. Some despise it for supposedly reducing everything to money and selfishness. (The theory was derided as "American" by some French authors,[19] even though it has its roots in France.) In fact, it is perfectly rational to want to live a life of self-denial, wishing only to reduce the suffering of others. Altruism simply means placing substantial weight on the well-being of others in evaluating any given future.

Another set of objections has to do with the difficulty of obtaining the necessary probabilities and utility values and multiplying them together to calculate expected utilities. These objections are simply confusing two different things: choosing the rational action and choosing it *by calculating expected utilities*. For example, if you try to poke your eyeball with your finger, your eyelid closes to protect your eye; this is rational, but no expected-utility calculations are involved. Or suppose you are riding a bicycle downhill with no brakes and have a choice between crashing into one concrete wall at ten miles per hour or another, farther down the hill, at twenty miles per hour; which would you prefer? If you chose ten miles per hour, congratulations! Did you calculate expected utilities? Probably not. But the choice of ten miles per hour is still rational. This follows from two basic assumptions: first, you prefer less severe injuries to more severe injuries, and second, for any given level of injuries, increasing the speed of collision increases the probability of exceeding that level. From these two assumptions it follows mathematically—without considering any numbers at all—that crashing at ten miles per hour has higher expected utility than crashing at twenty.[20] In summary, maximizing expected utility may not require calculating any expectations or any utilities. It's a purely *external* description of a rational entity.

Another critique of the theory of rationality lies in the identification of the locus of decision making. That is, what things count as agents? It might seem obvious that humans are agents, but what about families, tribes, corporations, cultures, and nation-states? If we examine social insects such as ants, does it make sense to consider a single ant as an intelligent agent, or does the intelligence really lie in the colony as a whole, with a kind of composite brain made up of multiple ant brains and bodies that are interconnected by pheromone signaling instead of electrical signaling? From an evolutionary point of view, this may be a more productive way of thinking about ants, since the ants in a given colony are typically closely related. As individuals, ants and other social insects seem to lack an instinct for self-preservation as distinct from colony preservation: they will always throw themselves into battle against invaders, even at suicidal odds. Yet sometimes humans will do the same even to defend unrelated humans; it is as if the species benefits from the presence of some fraction of individuals who are willing to sacrifice themselves in battle, or to go off on wild, speculative voyages of exploration, or to nurture the offspring of others. In such cases, an analysis of rationality that focuses entirely on the individual is clearly missing something essential.

The other principal objections to utility theory are empirical— that is, they are based on experimental evidence suggesting that humans are irrational. We fail to conform to the axioms in systematic ways.[21] It is not my purpose here to defend utility theory as a formal model of human behavior. Indeed, humans cannot possibly behave rationally. Our preferences extend over the whole of our own future lives, the lives of our children and grandchildren, and the lives of others, living now or in the future. Yet we cannot even play the right moves on the chessboard, a tiny, simple place with well-defined rules and a very short horizon. This is not because our *preferences* are irrational but because of the *complexity* of the decision problem. A great deal of our cognitive structure is there to compensate for the mismatch between our small, slow brains and the incomprehensibly huge complexity of the decision problem that we face all the time.

So, while it would be quite unreasonable to base a theory of beneficial AI on an assumption that humans are rational, it's quite reasonable to suppose that an adult human has roughly consistent preferences over future lives. That is, *if you were somehow able to watch two movies, each describing in sufficient detail and breadth a future life you might lead, such that each constitutes a virtual experience, you could say which you prefer, or express indifference.*[22]

This claim is perhaps stronger than necessary, if our only goal is to make sure that sufficiently intelligent machines are not catastrophic for the human race. The very notion of *catastrophe* entails a definitely-not-preferred life. For catastrophe avoidance, then, we need claim only that adult humans can recognize a catastrophic future when it is spelled out in great detail. Of course, human preferences have a much more fine-grained and, presumably, ascertainable structure than just "non-catastrophes are better than catastrophes."

A theory of beneficial AI can, in fact, accommodate inconsistency in human preferences, but the inconsistent part of your preferences can never be satisfied and there's nothing AI can do to help. Suppose, for example, that your preferences for pizza violate the axiom of transitivity:

> ROBOT: Welcome home! Want some pineapple pizza?
>
> YOU: No, you should know I prefer plain pizza to pineapple.
>
> ROBOT: OK, one plain pizza coming up!
>
> YOU: No thanks, I like sausage pizza better.
>
> ROBOT: So sorry, one sausage pizza!
>
> YOU: Actually, I prefer pineapple to sausage.
>
> ROBOT: My mistake, pineapple it is!
>
> YOU: I already said I like plain better than pineapple.

There is no pizza the robot can serve that will make you happy because there's always another pizza you would prefer to have. A robot can satisfy only the consistent part of your preferences—for example, let's say you prefer all three kinds of pizza to no pizza at all. In that case, a helpful robot could give you any one of the three pizzas, thereby satisfying your preference to avoid "no pizza" while leaving you to contemplate your annoyingly inconsistent pizza topping preferences at leisure.

Rationality for two

The basic idea that a rational agent acts so as to maximize expected utility is simple enough, even if actually doing it is impossibly complex. The theory applies, however, only in the case of a single agent acting alone. With more than one agent, the notion that it's possible—at least in principle—to assign probabilities to the different outcomes of one's actions becomes problematic. The reason is that now there's a part of the world—the other agent—that is trying to second-guess what action you're going to do, and vice versa, so it's not obvious how to assign probabilities to how that part of the world is going to behave. And without probabilities, the definition of rational action as maximizing expected utility isn't applicable.

As soon as someone else comes along, then, an agent will need some other way to make rational decisions. This is where *game theory* comes in. Despite its name, game theory isn't necessarily about games in the usual sense; it's a general attempt to extend the notion of rationality to situations with multiple agents. This is obviously important for our purposes, because we aren't planning (yet) to build robots that live on uninhabited planets in other star systems; we're going to put the robots in our world, which is inhabited by us.

To make it clear why we need game theory, let's look at a simple example: Alice and Bob playing soccer in the back garden (figure 3). Alice is about to take a penalty kick and Bob is in goal. Alice is going to shoot

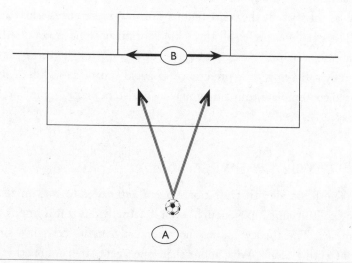

FIGURE 3: Alice about to take a penalty kick against Bob.

to Bob's left or to his right. Because she is right-footed, it's a little bit easier and more accurate for Alice to shoot to Bob's right. Because Alice has a ferocious shot, Bob knows he has to dive one way or the other right away—he won't have time to wait and see which way the ball is going. Bob could reason like this: "Alice has a better chance of scoring if she shoots to my right, because she's right-footed, so she'll choose that, so I'll dive right." But Alice is no fool and can imagine Bob thinking this way, in which case she will shoot to Bob's left. But Bob is no fool and can imagine Alice thinking this way, in which case he will dive to his left. But Alice is no fool and can imagine Bob thinking this way. . . . OK, you get the idea. Put another way: if there is a rational choice for Alice, Bob can figure it out too, anticipate it, and stop Alice from scoring, so the choice couldn't have been rational in the first place.

As early as 1713—once again, in the analysis of gambling games—a solution was found to this conundrum.[23] The trick is not to choose any one action but to choose a *randomized strategy*. For example, Alice can choose the strategy "shoot to Bob's right with probability 55 percent and shoot to his left with probability 45 percent." Bob could choose "dive right with probability 60 percent and left with probability 40 percent." Each mentally tosses a suitably biased coin just before acting, so they don't give away their intentions. By acting *unpredictably*, Alice and Bob avoid the contradictions of the preceding paragraph. Even if Bob works out what Alice's randomized strategy is, there's not much he can do about it without a crystal ball.

The next question is, What *should* the probabilities be? Is Alice's choice of 55 percent–45 percent rational? The specific values depend on how much more accurate Alice is when shooting to Bob's right, how good Bob is at saving the shot when he dives the right way, and so on. (See the notes for the complete analysis.[24]) The general criterion is very simple, however:

1. Alice's strategy is the best she can devise, assuming that Bob's is fixed.

2. Bob's strategy is the best he can devise, assuming that Alice's is fixed.

If both conditions are satisfied, we say that the strategies are in equilibrium. This kind of equilibrium is called a *Nash equilibrium* in honor of John Nash, who, in 1950 at the age of twenty-two, proved that such an equilibrium exists for any number of agents with any rational preferences and no matter what the rules of the game might be. After several decades' struggle with schizophrenia, Nash eventually recovered and was awarded the Nobel Memorial Prize in Economics for this work in 1994.

For Alice and Bob's soccer game, there is only one equilibrium. In other cases, there may be several, so the concept of Nash equilibria, unlike that of expected-utility decisions, does not always lead to a unique recommendation for how to behave.

Worse still, there are situations in which the Nash equilibrium seems to lead to highly undesirable outcomes. One such case is the famous *prisoner's dilemma*, so named by Nash's PhD adviser, Albert Tucker, in 1950.[25] The game is an abstract model of those all-too-common real-world situations where mutual cooperation would be better for all concerned but people nonetheless choose mutual destruction.

The prisoner's dilemma works as follows: Alice and Bob are suspects in a crime and are being interrogated separately. Each has a choice: to confess to the police and rat on his or her accomplice, or to refuse to talk.[26] If both refuse, they are convicted on a lesser charge and serve two years; if both confess, they are convicted on a more serious charge and serve ten years; if one confesses and the other refuses, the one who confesses goes free and the accomplice serves twenty years.

Now, Alice reasons as follows: "If Bob is going to confess, then I should confess too (ten years is better than twenty); if he is going to refuse, then I should confess (going free is better than spending two years in prison); so either way, I should confess." Bob reasons the same way. Thus, they both end up confessing to their crimes and serving ten years, even though by jointly refusing they could have served only two years. The problem is that joint refusal isn't a Nash equilibrium, because each has an incentive to defect and go free by confessing.

Note that Alice could have reasoned as follows: "Whatever reasoning I do, Bob will also do. So we'll end up choosing the same thing. Since joint refusal is better than joint confession, we should refuse." This form of reasoning acknowledges that, as rational agents, Alice and Bob will make choices that are correlated rather than independent. It's just one of many approaches that game theorists have tried in their efforts to obtain less depressing solutions to the prisoner's dilemma.[27]

Another famous example of an undesirable equilibrium is the *tragedy of the commons*, first analyzed in 1833 by the English economist William Lloyd[28] but named, and brought to global attention, by the ecologist Garrett Hardin in 1968.[29] The tragedy arises when several people can consume a shared resource—such as common grazing land or fish stocks—that replenishes itself slowly. Absent any social or legal constraints, the only Nash equilibrium among selfish (non-altruistic) agents is for each to consume as much as possible, leading to rapid collapse of the resource. The ideal solution, where everyone shares the resource such that the total consumption is sustainable, is not an equilibrium because each individual has an incentive to cheat and take more than their fair share—imposing the costs on others. In practice, of course, humans do sometimes avoid this tragedy by setting up mechanisms such as quotas and punishments or pricing schemes. They can do this because they are not limited to deciding how much to consume; they can also decide to *communicate*. By enlarging the decision problem in this way, we find solutions that are better for everyone.

These examples, and many others, illustrate the fact that extending the theory of rational decisions to multiple agents produces many interesting and complex behaviors. It's also extremely important because, as should be obvious, there is more than one human being. And soon there will be intelligent machines too. Needless to say, we have to achieve mutual cooperation, resulting in benefit to humans, rather than mutual destruction.

Computers

Having a reasonable definition of intelligence is the first ingredient in creating intelligent machines. The second ingredient is a machine in which that definition can be realized. For reasons that will soon become obvious, that machine is a computer. It *could* have been something different—for example, we might have tried to make intelligent machines out of complex chemical reactions or by hijacking biological cells[30]—but devices built for computation, from the very earliest mechanical calculators onwards, have always seemed to their inventors to be the natural home for intelligence.

We are so used to computers now that we barely notice their utterly incredible powers. If you have a laptop or a desktop or a smart phone, look at it: a small box, with a way to type characters. Just by typing, you can create programs that turn the box into something new, perhaps something that magically synthesizes moving images of oceangoing ships hitting icebergs or alien planets with tall blue people; type some more, and it translates English into Chinese; type some more, and it listens and speaks; type some more, and it defeats the world chess champion.

This ability of a single box to carry out any process that you can imagine is called *universality*, a concept first introduced by Alan Turing in 1936.[31] Universality means that we do not need separate machines for arithmetic, machine translation, chess, speech understanding, or animation: one machine does it all. Your laptop is essentially identical to the vast server farms run by the world's largest IT companies—even those equipped with fancy, special-purpose tensor processing units for machine learning. It's also essentially identical to all future computing devices yet to be invented. The laptop can do exactly the same tasks, provided it has enough memory; it just takes a lot longer.

Turing's paper introducing universality was one of the most important ever written. In it, he described a simple computing device that could accept as input the description of any other computing device, together with that second device's input, and, by simulating the operation of the second device on its input, produce the same output that the second device would have produced. We now call this first device a *universal Turing machine*. To prove its universality, Turing introduced precise definitions for two new kinds of mathematical objects: machines and programs. Together, the machine and program define a sequence of events—specifically, a sequence of state changes in the machine and its memory.

In the history of mathematics, new kinds of objects occur quite rarely. Mathematics began with numbers at the dawn of recorded history. Then, around 2000 BCE, ancient Egyptians and Babylonians worked with geometric objects (points, lines, angles, areas, and so on). Chinese mathematicians introduced matrices during the first millennium BCE, while sets as mathematical objects arrived only in the nineteenth century. Turing's new objects—machines and programs— are perhaps the most powerful mathematical objects ever invented. It is ironic that the field of mathematics largely failed to recognize this, and from the 1940s onwards, computers and computation have been the province of engineering departments in most major universities.

The field that emerged—computer science—exploded over the next seventy years, producing a vast array of new concepts, designs, methods, and applications, as well as seven of the eight most valuable companies in the world.

The central concept in computer science is that of an *algorithm*, which is a precisely specified method for computing something. Algorithms are, by now, familiar parts of everyday life: a square-root

algorithm in a pocket calculator receives a number as input and returns the square root of that number as output; a chess-playing algorithm takes a chess position and returns a move; a route-finding
algorithm takes a start location, a goal location, and a street map and
returns the fastest route from start to goal. Algorithms can be described in English or in mathematical notation, but to be implemented
they must be coded as programs in a *programming language*. More
complex algorithms can be built by using simpler ones as building
blocks called *subroutines*—for example, a self-driving car might use a
route-finding algorithm as a subroutine so that it knows where to go.
In this way, software systems of immense complexity are built up,
layer by layer.

Computer hardware matters because faster computers with more
memory allow algorithms to run more quickly and to handle more
information. Progress in this area is well known but still mind-
boggling. The first commercial electronic programmable computer,
the Ferranti Mark I, could execute about a thousand (10^3) instructions
per second and had about a thousand bytes of main memory. The fastest computer as of early 2019, the Summit machine at the Oak Ridge
National Laboratory in Tennessee, executes about 10^{18} instructions
per second (a thousand trillion times faster) and has 2.5×10^{17} bytes of
memory (250 trillion times more). This progress has resulted from
advances in electronic devices and even in the underlying physics, allowing an incredible degree of miniaturization.

Although comparisons between computers and brains are not especially meaningful, the numbers for Summit slightly exceed the raw
capacity of the human brain, which, as noted previously, has about
10^{15} synapses and a "cycle time" of about one hundredth of a second,
for a theoretical maximum of about 10^{17} "operations" per second. The
biggest difference is power consumption: Summit uses about a million
times more power.

Moore's law, an empirical observation that the number of electronic
components on a chip doubles every two years, is expected to continue

until 2025 or so, although at a slightly slower rate. For some years, speeds have been limited by the large amount of heat generated by the fast switching of silicon transistors; moreover, circuit sizes cannot get much smaller because the wires and connectors are (as of 2019) no more than twenty-five atoms wide and five to ten atoms thick. Beyond 2025, we will need to use more exotic physical phenomena—including negative capacitance devices,[32] single-atom transistors, graphene nano-tubes, and photonics—to keep Moore's law (or its successor) going.

Instead of just speeding up general-purpose computers, another possibility is to build special-purpose devices that are customized to perform just one class of computations. For example, Google's tensor processing units (TPUs) are designed to perform the calculations required for certain machine learning algorithms. One TPU pod (2018 version) performs roughly 10^{17} calculations per second—nearly as much as the Summit machine—but uses about one hundred times less power and is one hundred times smaller. Even if the underlying chip technology remains roughly constant, these kinds of machines can simply be made larger and larger to provide vast quantities of raw computational power for AI systems.

Quantum computation is a different kettle of fish. It uses the strange properties of quantum-mechanical wave functions to achieve something remarkable: with twice the amount of quantum hardware, you can do *more than twice* the amount of computation! Very roughly, it works like this:[33] Suppose you have a tiny physical device that stores a quantum bit, or qubit. A qubit has two possible states, 0 and 1. Whereas in classical physics the qubit device has to be in one of the two states, in quantum physics the *wave function* that carries information about the qubit says that it is in both states simultaneously. If you have two qubits, there are four possible joint states: 00, 01, 10, and 11. If the wave function is coherently entangled across the two qubits, meaning that no other physical processes are there to mess it up, then the two qubits are in all four states simultaneously. Moreover, if the two qubits are connected into a quantum circuit that performs some calculation, then the calculation proceeds with all four states simultaneously. With three qubits, you get eight states processed simultaneously, and so on. Now, there are some physical limitations so that the amount of work that gets done is less than exponential in the number of qubits,[34] but we know that there are important problems for which quantum computation is provably more efficient than any classical computer.

As of 2019, there are experimental prototypes of small quantum processors in operation with a few tens of qubits, but there are no interesting computing tasks for which a quantum processor is faster than a classical computer. The main difficulty is decoherence—processes such as thermal noise that mess up the coherence of the multi-qubit wave function. Quantum scientists hope to solve the decoherence problem by introducing error correction circuitry, so that any error that occurs in the computation is quickly detected and corrected by a kind of voting process. Unfortunately, error-correcting systems require far more qubits to do the same work: while a quantum machine with a few hundred perfect qubits would be very powerful compared to existing classical computers, we will probably need a few million error-correcting qubits to actually realize those computations. Going from a few tens to a few million qubits will take quite a few years. If, eventually, we get there, that would completely change the picture of what we can do by sheer brute-force computation.[35] Rather than waiting for real conceptual advances in AI, we might be able to use the raw power of quantum computation to bypass some of the barriers faced by current "unintelligent" algorithms.

The limits of computation

Even in the 1950s, computers were described in the popular press as "super-brains" that were "faster than Einstein." So can we say now, finally, that computers are as powerful as the human brain? No. Focusing on raw computing power misses the point entirely. Speed alone won't give us AI. Running a poorly designed algorithm on a faster computer doesn't make the algorithm better; it just means you get the wrong answer more quickly. (And with more data there are more opportunities for wrong answers!) The principal effect of faster machines has been to make the time for experimentation shorter, so that research can progress more quickly. It's not hardware that is holding AI back; it's software. We don't yet know how to make a machine really intelligent—even if it were the size of the universe.

Suppose, however, that we do manage to develop the right kind of AI software. Are there any limits placed by physics on how powerful a computer can be? Will those limits prevent us from having enough computing power to create real AI? The answers seem to be yes, there are limits, and no, there isn't a ghost of a chance that the limits will prevent us from creating real AI. MIT physicist Seth Lloyd has estimated the limits for a laptop-sized computer, based on considerations from quantum theory and entropy.[36] The numbers would raise even Carl Sagan's eyebrows: 10^{51} operations per second and 10^{30} bytes of memory, or approximately a billion trillion trillion times faster and four trillion times more memory than Summit—which, as noted previously, has more raw power than the human brain. Thus, when one hears suggestions that the human mind represents an upper limit on what is physically achievable in our universe,[37] one should at least ask for further clarification.

Besides limits imposed by physics, there are other limits on the abilities of computers that originate in the work of computer scientists. Turing himself proved that some problems are *undecidable* by any computer: the problem is well defined, there is an answer, but there cannot exist an algorithm that always finds that answer. He gave the example of what became known as the *halting problem*: Can an algorithm decide if a given program has an "infinite loop" that prevents it from ever finishing?[38]

Turing's proof that no algorithm can solve the halting problem[39] is incredibly important for the foundations of mathematics, but it seems to have no bearing on the issue of whether computers can be intelligent. One reason for this claim is that the same basic limitation seems to apply to the human brain. Once you start asking a human brain to perform an exact simulation of itself simulating itself simulating itself, and so on, you're bound to run into difficulties. I, for one, have never worried about my inability to do this.

Focusing on decidable problems, then, seems not to place any real restrictions on AI. It turns out, however, that decidable doesn't mean easy. Computer scientists spend a lot of time thinking about the *complexity* of problems, that is, the question of how much computation is needed to solve a problem by the most efficient method. Here's an easy problem: given a list of a thousand numbers, find the biggest number. If it takes one second to check each number, then it takes a thousand seconds to solve this problem by the obvious method of checking each in turn and keeping track of the biggest. Is there a faster method? No, because if a method *didn't* check some number in the list, that number might be the biggest, and the method would fail. So, the time to find the largest element is proportional to the size of the list. A computer scientist would say the problem has linear complexity, meaning that it's very easy; then she would look for something more interesting to work on.

What gets theoretical computer scientists excited is the fact that many problems appear[40] to have *exponential* complexity in the worst case. This means two things: first, all the algorithms we know about require exponential time—that is, an amount of time exponential in the size of the input—to solve at least some problem instances; second, theoretical computer scientists are pretty sure that more efficient algorithms do not exist.

Exponential growth in difficulty means that problems may be solvable in theory (that is, they are certainly decidable) but sometimes unsolvable in practice; we call such problems *intractable*. An example is the problem of deciding whether a given map can be colored with just three colors, so that no two adjacent regions have the same color.

(It is well known that coloring with four different colors is always possible.) With a million regions, it may be that there are some cases (not all, but some) that require something like 2^{1000} computational steps to find the answer, which means about 10^{275} years on the Summit supercomputer or a mere 10^{242} years on Seth Lloyd's ultimate-physics laptop. The age of the universe, about 10^{10} years, is a tiny blip compared to this.

Does the existence of intractable problems give us any reason to think that computers cannot be as intelligent as humans? No. There is no reason to suppose that humans can solve intractable problems either. Quantum computation helps a bit (whether in machines or brains), but not enough to change the basic conclusion.

Complexity means that the real-world decision problem—the problem of deciding what to do right now, at every instant in one's life—is so difficult that neither humans nor computers will ever come close to finding perfect solutions.

This has two consequences: first, we expect that, most of the time, real-world decisions will be at best halfway decent and certainly far from optimal; second, we expect that a great deal of the *mental architecture* of humans and computers—the way their decision processes actually operate—will be designed to overcome complexity to the extent possible—that is, to make it possible to find even halfway decent answers despite the overwhelming complexity of the world. Finally, we expect that the first two consequences will remain true no matter how intelligent and powerful some future machine may be. The machine may be far more capable than us, but it will still be far from perfectly rational.

Intelligent Computers

The development of logic by Aristotle and others made available precise rules for rational thought, but we do not know whether Aristotle ever contemplated the possibility of machines that implemented these rules. In the thirteenth century, the influential Catalan philosopher, seducer, and mystic Ramon Llull came much closer: he actually made paper wheels inscribed with symbols, by means of which he could generate logical combinations of assertions. The great seventeenth-century French mathematician Blaise Pascal was the first to develop a real and practical mechanical calculator. Although it could only add and subtract and was used mainly in his father's tax-collecting office, it led Pascal to write, "The arithmetical machine produces effects which appear nearer to thought than all the actions of animals."

Technology took a dramatic leap forward in the nineteenth century when the British mathematician and inventor Charles Babbage designed the Analytical Engine, a programmable universal machine in the sense defined later by Turing. He was helped in his work by Ada, Countess of Lovelace, daughter of the romantic poet and adventurer Lord Byron. Whereas Babbage hoped to use the Analytical Engine to compute accurate mathematical and astronomical tables, Lovelace understood its true potential,[41] describing it in 1842 as "a thinking or . . . a reasoning machine" that could reason about "all subjects in the universe." So, the basic conceptual elements for creating AI were in place! From that point, surely, AI would be just a matter of time. . . .

A long time, unfortunately—the Analytical Engine was never built, and Lovelace's ideas were largely forgotten. With Turing's theoretical work in 1936 and the subsequent impetus of World War II, universal computing machines were finally realized in the 1940s. Thoughts about creating intelligence followed immediately. Turing's 1950 paper, "Computing Machinery and Intelligence,"[42] is the best known of many early works on the possibility of intelligent machines. Skeptics were already asserting that machines would never be able to do X, for almost any X you could think of, and Turing refuted those assertions. He also proposed an operational test for intelligence, called the *imitation game*, which subsequently (in simplified form) became known as the *Turing test*. The test measures the *behavior* of the machine—specifically, its ability to fool a human interrogator into thinking that it is human.

The imitation game serves a specific role in Turing's paper—namely as a thought experiment to deflect skeptics who supposed that machines could not think in the right way, for the right reasons, with the right kind of awareness. Turing hoped to redirect the argument towards the issue of whether a machine could behave in a certain way; and if it did—if it was able, say, to discourse sensibly on Shakespeare's sonnets and their meanings—then skepticism about AI could not really be sustained. Contrary to common interpretations, I doubt that the test was intended as a true definition of intelligence, in the sense that a machine is intelligent if and only if it passes the Turing test. Indeed, Turing wrote, "May not machines carry out something which ought to be described as thinking but which is very different from what a man does?" Another reason not to view the test as a definition for AI is that it's a terrible definition to work with. And for that reason, mainstream AI researchers have expended almost no effort to pass the Turing test.

The Turing test is not useful for AI because it's an informal and highly contingent definition: it depends on the enormously complicated and largely unknown characteristics of the human mind, which derive from both biology and culture. There is no way to "unpack" the definition and work back from it to create machines that will provably pass the test. Instead, AI has focused on rational behavior, just as described previously: a machine is intelligent to the extent that what it does is likely to achieve what it wants, given what it has perceived.

Initially, like Aristotle, AI researchers identified "what it wants" with a goal that is either satisfied or not. These goals could be in toy worlds like the 15-puzzle, where the goal is to get all the numbered tiles lined up in order from 1 to 15 in a little (simulated) square tray; or they might be in real, physical environments: in the early 1970s, the Shakey robot at SRI in California was pushing large blocks into desired configurations, and Freddy at the University of Edinburgh was assembling a wooden boat from its component pieces. All this work was done using logical problem-solvers and planning systems to construct and execute guaranteed plans to achieve goals.[43]

By the 1980s, it was clear that logical reasoning alone could not suffice, because, as noted previously, there is no plan that is *guaranteed* to get you to the airport. Logic requires certainty, and the real world simply doesn't provide it. Meanwhile, the Israeli-American computer scientist Judea Pearl, who went on to win the 2011 Turing Award, had been working on methods for uncertain reasoning based in probability theory.[44] AI researchers gradually accepted Pearl's ideas; they adopted the tools of probability theory and utility theory and thereby connected AI to other fields such as statistics, control theory, economics, and operations research. This change marked the beginning of what some observers call *modern AI*.

Agents and environments

The central concept of modern AI is the *intelligent agent*— something that perceives and acts. The agent is a process occurring over time, in the sense that a stream of perceptual inputs is converted into a stream of actions. For example, suppose the agent in question is a self-driving taxi taking me to the airport. Its inputs might include eight RGB cameras operating at thirty frames per second; each frame consists of perhaps 7.5 million pixels, each with an image intensity value in each of three color channels, for a total of more than five gigabytes per second. (The flow of data from the two hundred million photoreceptors in the retina is even larger, which partially explains why vision occupies such a large fraction of the human brain.) The taxi also gets data from an accelerometer one hundred times per second, as well as GPS data. This incredible flood of raw data is transformed by the simply gargantuan computing power of billions of transistors (or neurons) into smooth, competent driving behavior. The taxi's actions include the electronic signals sent to the steering wheel, brakes, and accelerator, twenty times per second. (For an experienced human driver, most of this maelstrom of activity is unconscious: you may be aware only of making decisions such as "overtake this slow truck" or "stop for gas," but your eyes, brain, nerves, and muscles are still doing all the other stuff.) For a chess program, the inputs are mostly just the clock ticks, with the occasional notification of the opponent's move and the new board state, while the actions are mostly doing nothing while the program is thinking, and occasionally choosing a move and notifying the opponent. For a personal digital assistant, or PDA, such as Siri or Cortana, the inputs include not just the acoustic signal from the microphone (sampled forty-eight thousand times per second) and input from the touch screen but also the content of each Web page that it accesses, while the actions include both speaking and displaying material on the screen.

The way we build intelligent agents depends on the nature of the problem we face. This, in turn, depends on three things: first, the nature of the environment the agent will operate in—a chessboard is a very different place from a crowded freeway or a mobile phone; second, the observations and actions that connect the agent to the environment—for example, Siri might or might not have access to the phone's camera so that it can see; and third, the agent's objective—teaching the opponent to play better chess is a very different task from winning the game.

To give just one example of how the design of the agent depends on these things: If the objective is to win the game, a chess program need consider only the current board state and does not need any memory of past events.[45] The chess tutor, on the other hand, should continually update its model of which aspects of chess the pupil does or does not understand so that it can provide useful advice. In other words, for the chess tutor, the pupil's mind is a relevant part of the environment. Moreover, unlike the board, it is a part of the environment that is not directly observable.

The characteristics of problems that influence the design of agents include at least the following:[46]

- whether the environment is fully observable (as in chess, where the inputs provide direct access to all the relevant aspects of the current state of the environment) or partially observable (as in driving, where one's field of view is limited, vehicles are opaque, and other drivers' intentions are mysterious);
- whether the environment and actions are discrete (as in chess) or effectively continuous (as in driving);
- whether the environment contains other agents (as in chess and driving) or not (as in finding the shortest routes on a map);
- whether the outcomes of actions, as specified by the "rules" or "physics" of the environment, are predictable (as in chess) or unpredictable (as in traffic and weather), and whether those rules are known or unknown;
- whether the environment is dynamically changing, so that the time to make decisions is tightly constrained (as in driving) or not (as in tax strategy optimization);
- the length of the horizon over which decision quality is measured according to the objective—this may be very short (as in emergency braking), of intermediate duration (as in chess, where a game lasts up to about one hundred moves), or very long (as in driving me to the airport, which might take hundreds of thousands of decision cycles if the taxi is deciding one hundred times per second).

As one can imagine, these characteristics give rise to a bewildering variety of problem types. Just multiplying the choices listed above gives 192 types. One can find real-world problem instances for all the types. Some types are typically studied in areas outside AI—for example, designing an autopilot that maintains level flight is a short-horizon, continuous, dynamic problem that is usually studied in the field of control theory.

Obviously some problem types are easier than others. AI has made a lot of progress on problems such as board games and puzzles that are observable, discrete, deterministic, and have known rules. For the easier problem types, AI researchers have developed fairly general and effective algorithms and a solid theoretical understanding; often, machines exceed human performance on these kinds of problems. We can tell that an algorithm is general because we have mathematical proofs that it gives optimal or near-optimal results with reasonable computational complexity across an entire class of problems, and because it works well in practice on those kinds of problems without needing any problem-specific modifications.

Video games such as StarCraft are quite a bit harder than board games: they involve hundreds of moving parts and time horizons of thousands of steps, and the board is only partially visible at any given time. At each point, a player might have a choice of at least 10^{50} moves, compared to about 10^2 in Go.[47] On the other hand, the rules are known and the world is discrete with only a few types of objects. As of early 2019, machines are as good as some professional StarCraft players but not yet ready to challenge the very best humans.[48] More important, it took a fair amount of problem-specific effort to reach that point; general-purpose methods are not quite ready for StarCraft.

Problems such as running a government or teaching molecular biology are *much* harder. They have complex, mostly unobservable environments (the state of a whole country, or the state of a student's mind), far more objects and types of objects, no clear definition of what the actions are, mostly unknown rules, a great deal of uncertainty, and very long time scales. We have ideas and off-the-shelf tools that address each of these characteristics separately but, as yet, no general methods that cope with all the characteristics simultaneously. When we build AI systems for these kinds of tasks, they tend to require a great deal of problem-specific engineering and are often very brittle.

Progress towards generality occurs when we devise methods that are effective for harder problems within a given type or methods that require fewer and weaker assumptions so they are applicable to more problems. General-purpose AI would be a method that is applicable across all problem types and works effectively for large and difficult instances while making very few assumptions. That's the ultimate goal of AI research: a system that needs no problem-specific engineering and can simply be asked to teach a molecular biology class or run a government. It would learn what it needs to learn from all the available resources, ask questions when necessary, and begin formulating and executing plans that work.

Such a general-purpose method does not yet exist, but we are moving closer. Perhaps surprisingly, a lot of this progress towards general AI results from research that isn't about building scary, general-purpose AI systems. It comes from research on *tool AI* or *narrow AI*, meaning nice, safe, boring AI systems designed for particular problems such as playing Go or recognizing handwritten digits. Research on this kind of AI is often thought to present no risk because it's problem-specific and nothing to do with general-purpose AI.

This belief results from a misunderstanding of what kind of work goes into these systems. In fact, research on tool AI can and often does produce progress towards general-purpose AI, particularly when it is done by researchers with good taste attacking problems that are beyond the capabilities of current general methods. Here, *good taste* means that the solution approach is not merely an ad hoc encoding of what an intelligent person would do in such-and-such situation but an attempt to provide the machine with the ability to figure out the solution for itself.

For example, when the AlphaGo team at Google DeepMind succeeded in creating their world-beating Go program, they did this *without really working on Go*. What I mean by this is that they didn't write a whole lot of Go-specific code saying what to do in different kinds of Go situations. They didn't design decision procedures that work only for Go. Instead, they made improvements to two fairly general-purpose techniques—lookahead search to make decisions and reinforcement learning to learn how to evaluate positions—so that they were sufficiently effective to play Go at a superhuman level. Those improvements are applicable to many other problems, including problems as far afield as robotics. Just to rub it in, a version of AlphaGo called AlphaZero recently learned to trounce AlphaGo at Go, and also to trounce Stockfish (the world's best chess program, far better than any human) and Elmo (the world's best shogi program, also better than any human). AlphaZero did all this in one day.[49]

There was also substantial progress towards general-purpose AI in research on recognizing handwritten digits in the 1990s. Yann LeCun's team at AT&T Labs didn't write special algorithms to recognize "8" by searching for curvy lines and loops; instead, they improved on existing neural network learning algorithms to produce *convolutional neural networks*. Those networks, in turn, exhibited effective character recognition after suitable training on labeled examples. The same algorithms can learn to recognize letters, shapes, stop signs, dogs, cats, and police cars. Under the headline of "deep learning," they have revolutionized speech recognition and visual object recognition. They are also one of the key components in AlphaZero as well as in most of the current self-driving car projects.

If you think about it, it's hardly surprising that progress towards general AI is going to occur in narrow-AI projects that address specific tasks; those tasks give AI researchers something to get their teeth into. (There's a reason people don't say, "Staring out the window is the mother of invention.") At the same time, it's important to understand how much progress has occurred and where the boundaries are. When AlphaGo defeated Lee Sedol and later all the other top Go players, many people assumed that because a machine had learned from scratch to beat the human race at a task known to be very difficult even for highly intelligent humans, it was the beginning of the end—just a matter of time before AI took over. Even some skeptics may have been convinced when AlphaZero won at chess and shogi as well as Go. But AlphaZero has hard limitations: it works only in the class of discrete, observable, two-player games with known rules. The approach simply won't work *at all* for driving, teaching, running a government, or taking over the world.

These sharp boundaries on machine competence mean that when people talk about "machine IQ" increasing rapidly and threatening to exceed human IQ, they are talking nonsense. To the extent that the concept of IQ makes sense when applied to humans, it's because human abilities tend to be correlated across a wide range of cognitive activities. Trying to assign an IQ to machines is like trying to get four-legged animals to compete in a human decathlon. True, horses can run fast and jump high, but they have a lot of trouble with pole-vaulting and throwing the discus.

Objectives and the standard model

Looking at an intelligent agent from the outside, what matters is the stream of actions it generates from the stream of inputs it receives. From the inside, the actions have to be chosen by an *agent program*. Humans are born with one agent program, so to speak, and that program learns over time to act reasonably successfully across a huge range of tasks. So far, that is not the case for AI: we don't know how to build one general-purpose AI program that does everything, so instead we build different types of agent programs for different types of problems. I will need to explain at least a tiny bit about how these different agent programs work; more detailed explanations are given in the appendices at the end of the book for those who are interested. (Pointers to particular appendices are given as superscripts like this[A] and this.[D]) The primary focus here is on how the standard model is instantiated in these various kinds of agents—in other words, how the objective is specified and communicated to the agent.

The simplest way to communicate an objective is in the form of a *goal*. When you get into your self-driving car and touch the "home" icon on the screen, the car takes this as its objective and proceeds to plan and execute a route. A state of the world either satisfies the goal (yes, I'm at home) or it doesn't (no, I don't live at the San Francisco Airport). In the classical period of AI research, before uncertainty became a primary issue in the 1980s, most AI research assumed a world that was fully observable and deterministic, and goals made sense as a way to specify objectives. Sometimes there is also a *cost function* to evaluate solutions, so an optimal solution is one that minimizes total cost while reaching the goal. For the car, this might be built in—perhaps the cost of a route is some fixed combination of the time and fuel consumption—or the human might have the option of specifying the trade-off between the two.

The key to achieving such objectives is the ability to "mentally simulate" the effects of possible actions, sometimes called *lookahead search*. Your self-driving car has an internal map, so it knows that driving east from San Francisco on the Bay Bridge gets you to Oakland. Algorithms originating in the 1960s[50] find optimal routes by looking ahead and searching through many possible action sequences.[A] These algorithms form a ubiquitous part of modern infrastructure: they provide not just driving directions but also airline travel solutions, robotic assembly, construction planning, and delivery logistics. With some modifications to handle the impertinent behavior of opponents, the same idea of lookahead applies to games such as tic-tac-toe, chess, and Go, where the goal is to win according to the game's particular definition of winning.

Lookahead algorithms are incredibly effective for their specific tasks, but they are not very flexible. For example, AlphaGo "knows" the rules of Go, but only in the sense that it has two subroutines, written in a traditional programming language such as C++: one sub-routine generates all the possible legal moves and the other encodes the goal, determining whether a given state is won or lost. For AlphaGo to play a different game, someone has to rewrite all this C++ code. Moreover, if you give it a new goal—say, visiting the exoplanet that orbits Proxima Centauri—it will explore billions of sequences of Go moves in a vain attempt to find a sequence that achieves the goal. It cannot look inside the C++ code and determine the obvious: no sequence of Go moves gets you to Proxima Centauri. AlphaGo's knowledge is essentially locked inside a black box.

In 1958, two years after his Dartmouth summer meeting had ini-tiated the field of artificial intelligence, John McCarthy proposed a much more general approach that opens up the black box: writing general-purpose reasoning programs that can absorb knowledge on any topic and reason with it to answer any answerable question.[51] One particular kind of reasoning would be practical reasoning of the kind suggested by Aristotle: "Doing actions A, B, C, . . . will achieve goal G." The goal could be anything at all: make sure the house is tidy be-fore I get home, win a game of chess without losing either of your knights, reduce my taxes by 50 percent, visit Proxima Centauri, and so on. McCarthy's new class of programs soon became known as knowledge-based systems.[52]

To make knowledge-based systems possible requires answering two questions. First, how can knowledge be stored in a computer? Second, how can a computer reason correctly with that knowledge to draw new conclusions? Fortunately, ancient Greek philosophers—particularly Aristotle—provided basic answers to these questions long before the advent of computers. In fact, it seems quite likely that, had Aristotle been given access to a computer (and some electricity, I sup-pose), he would have been an AI researcher. Aristotle's answer, reiter-ated by McCarthy, was to use formal logic[B] as the basis for knowledge and reasoning.

There are two kinds of logic that really matter in computer science. The first, called *propositional* or *Boolean logic*, was known to the Greeks as well as to ancient Chinese and Indian philosophers. It is the same language of AND gates, NOT gates, and so on that makes up the circuitry of computer chips. In a very literal sense, a modern CPU is just a very large mathematical expression—hundreds of millions of pages—written in the language of propositional logic. The second kind of logic, and the one that McCarthy proposed to use for AI, is called *first-order* logic.[B] The language of first-order logic is far more expressive than propositional logic, which means that there are things that can be expressed very easily in first-order logic that are painful or impossible to write in propositional logic. For example, the rules of Go take about a page in first-order logic but millions of pages in propositional logic. Similarly, we can easily express knowledge about chess, British citizenship, tax law, buying and selling, moving, painting, cooking, and many other aspects of our commonsense world.

In principle, then, the ability to reason with first-order logic gets us a long way towards general-purpose intelligence. In 1930, the brilliant Austrian logician Kurt Gödel had published his famous *completeness theorem*,[53] proving that there is an algorithm with the following property:[54]

> For *any collection of knowledge* and *any question* expressible in first-order logic, the algorithm will tell us the answer to the question if there is one.

This is a pretty incredible guarantee. It means, for example, that we can tell the system the rules of Go and it will tell us (if we wait long enough) whether there is an opening move that wins the game. We can tell it facts about local geography, and it will tell us the way to the airport. We can tell it facts about geometry and motion and utensils, and it will tell the robot how to lay the table for dinner. More

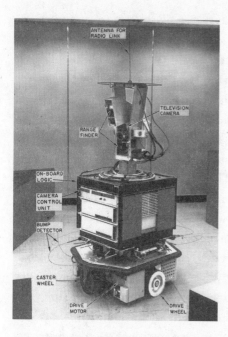

FIGURE 4: Shakey the robot, circa 1970. In the background are some of the objects that Shakey pushed around in its suite of rooms.

generally, given any achievable goal and sufficient knowledge of the effects of its actions, an agent can use the algorithm to construct a plan that it can execute to achieve the goal.

It must be said that Gödel did not actually provide an algorithm; he merely proved that one existed. In the early 1960s, real algorithms for logical reasoning began to appear,[55] and McCarthy's dream of generally intelligent systems based on logic seemed within reach. The first major mobile robot project in the world, SRI's Shakey project, was based on logical reasoning (see figure 4). Shakey received a goal from its human designers, used vision algorithms to create logical assertions describing the current situation, performed logical inference to derive a guaranteed plan to achieve the goal, and then executed the plan. Shakey was "living" proof that Aristotle's analysis of human cognition and action was at least partially correct.

Unfortunately, Aristotle's (and McCarthy's) analysis was far from being completely correct. The main problem is ignorance—not, I hasten to add, on the part of Aristotle or McCarthy, but on the part of all humans and machines, present and future. Very little of our knowledge is absolutely certain. In particular, we don't know very much about the future. Ignorance is just an insuperable problem for a purely logical system. If I ask, "Will I get to the airport on time, if I leave three hours before my flight?" or "Can I obtain a house by buying a winning lottery ticket and then buying the house with the proceeds?" the correct answer will be, in each case, "I don't know." The reason is that, for each question, both yes and no are logically possible. As a practical matter, one can never be absolutely certain of any empirical question unless the answer is already known.[56] Fortunately, certainty is completely unnecessary for action: we just need to know which action is best, not which action is certain to succeed.

Uncertainty means that the "purpose put into the machine" cannot, in general, be a precisely delineated goal, to be achieved at all costs. There is no longer such a thing as a "sequence of actions that achieves the goal," because any sequence of actions will have multiple possible outcomes, some of which won't achieve the goal. The likelihood of success really matters: leaving for the airport three hours in advance of your flight *may* mean that you won't miss the flight and buying a lottery ticket *may* mean that you'll win enough to buy a new house, but these are very different *mays*. Goals cannot be rescued by looking for plans that maximize the probability of achieving the goal. A plan that maximizes the probability of getting to the airport in time to catch a flight might involve leaving home days in advance, organizing an armed escort, lining up many alternative means of transport in case the others break down, and so on. Inevitably, one must take into account the relative desirabilities of different outcomes as well as their likelihoods.

Instead of a goal, then, we could use a utility function to describe the desirability of different outcomes or sequences of states. Often, the utility of a sequence of states is expressed as a sum of *rewards* for each of the states in the sequence. Given a purpose defined by a utility or reward function, the machine aims to produce behavior that maximizes its expected utility or expected sum of rewards, averaged over the possible outcomes weighted by their probabilities. Modern AI is partly a rebooting of McCarthy's dream, except with utilities and probabilities instead of goals and logic.

Pierre-Simon Laplace, the great French mathematician, wrote in 1814, "The theory of probabilities is just common sense reduced to calculus."[57] It was not until the 1980s, however, that a practical formal language and reasoning algorithms were developed for probabilistic knowledge. This was the language of *Bayesian networks*,[C] introduced by Judea Pearl. Roughly speaking, Bayesian networks are the probabilistic cousins of propositional logic. There are also probabilistic cousins of first-order logic, including Bayesian logic[58] and a wide variety of *probabilistic programming languages.*

Bayesian networks and Bayesian logic are named after the Reverend Thomas Bayes, a British clergyman whose lasting contribution to modern thought—now known as *Bayes' theorem*—was published in 1763, shortly after his death, by his friend Richard Price.[59] In its modern form, as suggested by Laplace, the theorem describes in a very simple way how a *prior* probability—the initial degree of belief one has in a set of possible hypotheses—becomes a *posterior* probability as a result of observing some evidence. As more new evidence arrives, the posterior becomes the new prior and the process of Bayesian updating repeats ad infinitum. This process is so fundamental that the modern idea of rationality as maximization of expected utility is sometimes called *Bayesian rationality*. It assumes that a rational agent has access to a posterior probability distribution over possible current states of the world, as well as over hypotheses about the future, based on all its past experience.

Researchers in operations research, control theory, and AI have also developed a variety of algorithms for decision making under uncertainty, some dating back to the 1950s. These so-called "dynamic programming" algorithms are the probabilistic cousins of lookahead search and planning and can generate optimal or near-optimal behavior for all sorts of practical problems in finance, logistics, transportation, and so on, where uncertainty plays a significant role.[c] The purpose is put into these machines in the form of a reward function, and the output is a *policy* that specifies an action for every possible state the agent could get itself into.

For complex problems such as backgammon and Go, where the number of states is enormous and the reward comes only at the end of the game, lookahead search won't work. Instead, AI researchers have developed a method called *reinforcement learning*, or RL for short. RL algorithms learn from direct experience of reward signals in the environment, much as a baby learns to stand up from the positive reward of being upright and the negative reward of falling over. As with dynamic programming algorithms, the purpose put into an RL algorithm is the reward function, and the algorithm learns an estimator for the value of states (or sometimes the value of actions). This estimator can be combined with relatively myopic lookahead search to generate highly competent behavior.

The first successful reinforcement learning system was Arthur Samuel's checkers program, which created a sensation when it was demonstrated on television in 1956. The program learned essentially from scratch, by playing against itself and observing the rewards of winning and losing.[60] In 1992, Gerry Tesauro applied the same idea to the game of backgammon, achieving world-champion-level play after 1,500,000 games.[61] Beginning in 2016, DeepMind's AlphaGo and its descendants used reinforcement learning and self-play to defeat the best human players at Go, chess, and shogi.

Reinforcement learning algorithms can also learn how to select actions based on raw perceptual input. For example, DeepMind's DQN system learned to play forty-nine different Atari video games entirely from scratch—including Pong, Freeway, and Space Invaders.[62] It used only the screen pixels as input and the game score as a reward signal. In most of the games, DQN learned to play better than a professional human player—despite the fact that DQN has no a priori notion of time, space, objects, motion, velocity, or shooting. It is quite hard to work out what DQN is actually doing, besides winning.

If a newborn baby learned to play dozens of video games at superhuman levels on its first day of life, or became world champion at Go, chess, and shogi, we might suspect demonic possession or alien intervention. Remember, however, that all these tasks are much simpler than the real world: they are fully observable, they involve short time horizons, and they have relatively small state spaces and simple, predictable rules. Relaxing any of these conditions means that the standard methods will fail.

Current research, on the other hand, is aimed precisely at going beyond standard methods so that AI systems can operate in larger classes of environments. On the day I wrote the preceding paragraph, for example, OpenAI announced that its team of five AI programs had learned to beat experienced human teams at the game Dota 2. (For the uninitiated, who include me: Dota 2 is an updated version of *Defense of the Ancients*, a real-time strategy game in the Warcraft family; it is currently the most lucrative and competitive e-sport, with prizes in the millions of dollars.) Dota 2 involves communication, teamwork, and quasi-continuous time and space. Games last for tens of thousands of time steps, and some degree of hierarchical organization of behavior seems to be essential. Bill Gates described the announcement as "a huge milestone in advancing artificial intelligence."[63] A few months later, an updated version of the program defeated the world's top professional Dota 2 team.[64]

Games such as Go and Dota 2 are a good testing ground for reinforcement learning methods because the reward function comes with the rules of the game. The real world is less convenient, however, and there have been dozens of cases in which faulty definitions of rewards led to weird and unanticipated behaviors.[65] Some are innocuous, like the simulated evolution system that was supposed to evolve fast-moving creatures but in fact produced creatures that were enormously tall and moved fast by falling over.[66] Others are less innocuous, like the social-media click-through optimizers that seem to be making a fine mess of our world.

The final category of agent program I will consider is the simplest: programs that connect perception directly to action, without any intermediate deliberation or reasoning. In AI, we call this kind of program a *reflex agent*—a reference to the low-level neural reflexes exhibited by humans and animals, which are not mediated by thought.[67] For example, the human blinking reflex connects the outputs of low-level processing circuits in the visual system directly to the motor area that controls the eyelids, so that any rapidly looming region in the visual field causes a hard blink. You can test it now by trying (not too hard) to poke yourself in the eye with your finger. We can think of this reflex system as a simple "rule" of the following form:

if *<rapidly looming region in visual field>* then *<blink>*.

The blinking reflex does not "know what it's doing": the objective (of shielding the eyeball from foreign objects) is nowhere represented; the knowledge (that a rapidly looming region corresponds to an object approaching the eye, and that an object approaching the eye might damage it) is nowhere represented. Thus, when the non-reflex part of you wants to put in eye drops, the reflex part still blinks.

Another familiar reflex is emergency braking—when the car in front stops unexpectedly or a pedestrian steps into the road. Quickly deciding whether braking is required is not easy: when a test vehicle in autonomous mode killed a pedestrian in 2018, Uber explained that "emergency braking maneuvers are not enabled while the vehicle is under computer control, to reduce the potential for erratic vehicle behavior."[68] Here, the human designer's objective is clear—don't kill pedestrians—but the agent's policy (had it been activated) implements it incorrectly. Again, the objective is not represented in the agent: no autonomous vehicle today knows that people don't like to be killed.

Reflex actions also play a role in more routine tasks such as staying in lane: as the car drifts ever so slightly out of the ideal lane position, a simple feedback control system can nudge the steering wheel in the opposite direction to correct the drift. The size of the nudge would depend on how far the car drifted. These kinds of control systems are usually designed to minimize the square of the tracking error added up over time. The designer derives a feedback control law that, under certain assumptions about speed and road curvature, approximately implements this minimization.[69] A similar system is operating all the time while you are standing up; if it were to stop working, you'd fall over within a few seconds. As with the blinking reflex, it's quite hard to turn this mechanism off and allow yourself to fall over.

Reflex agents, then, implement a designer's objective, but do not know what the objective is or why they are acting in a certain way. This means they cannot really make decisions for themselves; someone else, typically the human designer or perhaps the process of biological evolution, has to decide everything in advance. It is very hard to create a good reflex agent by manual programming except for very simple tasks such as tic-tac-toe or emergency braking. Even in those cases, the reflex agent is extremely inflexible and cannot change its behavior when circumstances indicate that the implemented policy is no longer appropriate.

One possible way to create more powerful reflex agents is through a process of learning from examples.[D] Rather than specifying a rule for how to behave, or supplying a reward function or a goal, a human can supply examples of decision problems along with the correct decision to make in each case. For example, we can create a French-to-English translation agent by supplying examples of French sentences along with the correct English translations. (Fortunately, the Canadian and EU parliaments generate millions of such examples every year.) Then a *supervised learning* algorithm processes the examples to produce a complex rule that takes any French sentence as input and produces an English translation. The current champion learning algorithm for machine translation is a form of so-called deep learning, and it produces a rule in the form of an artificial neural network with hundreds of layers and millions of parameters.[D] Other deep learning algorithms have turned out to be very good at classifying the objects in images and recognizing the words in a speech signal. Machine translation, speech recognition, and visual object recognition are three of the most important subfields in AI, which is why there has been so much excitement about the prospects for deep learning.

One can argue almost endlessly about whether deep learning will lead directly to human-level AI. My own view, which I will explain later, is that it falls far short of what is needed,[D] but for now let's focus on how such methods fit into the standard model of AI, where an algorithm optimizes a fixed objective. For deep learning, or indeed for any supervised learning algorithm, the "purpose put into the machine" is usually to maximize predictive accuracy—or, equivalently, to minimize error. That much seems obvious, but there are actually two ways to understand it, depending on the role that the learned rule is going to play in the overall system. The first role is a purely perceptual role: the network processes the sensory input and provides information to the rest of the system in the form of probability estimates for what it's perceiving. If it's an object recognition algorithm, maybe it says "70 percent probability it's a Norfolk terrier, 30 percent it's a Norwich terrier."[70] The rest of the system decides on an external action to take based on this information. This purely perceptual objective is unproblematic in the following sense: even a "safe" superintelligent AI system, as opposed to an "unsafe" one based on the standard model, needs to have its perception system as accurate and well calibrated as possible.

The problem comes when we move from a purely perceptual role to a decision-making role. For example, a trained network for recognizing objects might automatically generate labels for images on a

Web site or social-media account. Posting those labels is an action with consequences. Each labeling action requires an actual classification decision, and unless every decision is guaranteed to be perfect, the human designer must supply a *loss function* that spells out the cost of misclassifying an object of type A as an object of type B. And that's how Google had an unfortunate problem with gorillas. In 2015, a software engineer named Jacky Alciné complained on Twitter that the Google Photos image-labeling service had labeled him and his friend as gorillas.[71] While it is unclear how exactly this error occurred, it is almost certain that Google's machine learning algorithm was designed to minimize a fixed, definite loss function—moreover, one that assigned equal cost to any error. In other words, it assumed that the cost of misclassifying a person as a gorilla was the same as the cost of misclassifying a Norfolk terrier as a Norwich terrier. Clearly, this is not Google's (or their users') true loss function, as was illustrated by the public relations disaster that ensued.

Since there are thousands of possible image labels, there are millions of potentially distinct costs associated with misclassifying one category as another. Even if it had tried, Google would have found it very difficult to specify all these numbers up front. Instead, the right thing to do would be to acknowledge the uncertainty about the true misclassification costs and to design a learning and classification algorithm that was suitably sensitive to costs and uncertainty about costs. Such an algorithm might occasionally ask the Google designer questions such as "Which is worse, misclassifying a dog as a cat or misclassifying a person as an animal?" In addition, if there is significant uncertainty about misclassification costs, the algorithm might well refuse to label some images.

By early 2018, it was reported that Google Photos does refuse to classify a photo of a gorilla. Given a very clear image of a gorilla with two babies, it says, "Hmm . . . not seeing this clearly yet."[72]

I don't wish to suggest that AI's adoption of the standard model was a poor choice at the time. A great deal of brilliant work has gone into developing the various instantiations of the model in logical, probabilistic, and learning systems. Many of the resulting systems are very useful; as we will see in the next chapter, there is much more to come. On the other hand, we cannot continue to rely on our usual practice of ironing out the major errors in an objective function by trial and error: machines of increasing intelligence and increasingly global impact will not allow us that luxury.

HOW MIGHT AI PROGRESS IN THE FUTURE?

The Near Future

On May 3, 1997, a chess match began between Deep Blue, a chess computer built by IBM, and Garry Kasparov, the world chess champion and possibly the best human player in history. *Newsweek* billed the match as "The Brain's Last Stand." On May 11, with the match tied at 2½–2½, Deep Blue defeated Kasparov in the final game. The media went berserk. The market capitalization of IBM increased by $18 billion overnight. AI had, by all accounts, achieved a massive breakthrough.

From the point of view of AI research, the match represented no breakthrough at all. Deep Blue's victory, impressive as it was, merely continued a trend that had been visible for decades. The basic design for chess-playing algorithms was laid out in 1950 by Claude Shannon,[1] with major improvements in the early 1960s. After that, the chess ratings of the best programs improved steadily, mainly as a result of faster computers that allowed programs to look further ahead. In 1994,[2] Peter Norvig and I charted the numerical ratings of the best chess programs from 1965 onwards, on a scale where Kasparov's rating was 2805. The ratings started at 1400 in 1965 and improved in an almost perfect straight line for thirty years. Extrapolating the line forward from 1994 predicts that computers would be able to defeat Kasparov in 1997—exactly when it happened.

For AI researchers, then, the real breakthroughs happened thirty or forty years *before* Deep Blue burst into the public's consciousness. Similarly, deep convolutional networks existed, with all the mathematics fully worked out, more than twenty years before they began to create headlines.

The view of AI breakthroughs that the public gets from the media—stunning victories over humans, robots becoming citizens of Saudi Arabia, and so on—bears very little relation to what really happens in the world's research labs. Inside the lab, research involves a lot of thinking and talking and writing mathematical formulas on whiteboards. Ideas are constantly being generated, abandoned, and rediscovered. A good idea—a real breakthrough—will often go unnoticed at the time and may only later be understood as having provided the basis for a substantial advance in AI, perhaps when someone reinvents it at a more convenient time. Ideas are tried out, initially on simple problems to show that the basic intuitions are correct and then on harder problems to see how well they scale up. Often, an idea will fail by itself to provide a substantial improvement in capabilities, and it has to wait for another idea to come along so that the combination of the two can demonstrate value.

All this activity is completely invisible from the outside. In the world beyond the lab, AI becomes visible only when the gradual accumulation of ideas and the evidence for their validity crosses a threshold: the point where it becomes worthwhile to invest money and engineering effort to create a new commercial product or an impressive demonstration. Then the media announce that a breakthrough has occurred.

One can expect, then, that many other ideas that have been

gestating in the world's research labs will cross the threshold of commercial applicability over the next few years. This will happen more and more frequently as the rate of commercial investment increases and as the world becomes more and more receptive to applications of AI. This chapter provides a sampling of what we can see coming down the pipe.

Along the way, I'll mention some of the drawbacks of these technological advances. You will probably be able to think of many more, but don't worry. I'll get to those in the next chapter.

The AI ecosystem

In the beginning, the environment in which most computers operated was essentially formless and void: their only input came from punched cards and their only method of output was to print characters on a line printer. Perhaps for this reason, most researchers viewed intelligent machines as question-answerers; the view of machines as *agents* perceiving and acting in an environment did not become widespread until the 1980s.

The advent of the World Wide Web in the 1990s opened up a whole new universe for intelligent machines to play in. A new word, *softbot*, was coined to describe software "robots" that operate entirely in a software environment such as the Web. Softbots, or bots as they later became known, perceive Web pages and act by emitting sequences of characters, URLs, and so on.

AI companies mushroomed during the dot-com boom (1997–2000), providing core capabilities for search and e-commerce, including link analysis, recommendation systems, reputation systems, comparison shopping, and product categorization.

In the early 2000s, the widespread adoption of mobile phones with microphones, cameras, accelerometers, and GPS provided new access for AI systems to people's daily lives; "smart speakers" such as the Amazon Echo, Google Home, and Apple HomePod have completed this process.

By around 2008, the number of objects connected to the Internet exceeded the number of people connected to the Internet—a transition that some point to as the beginning of the Internet of Things (IoT). Those things include cars, home appliances, traffic lights, vending machines, thermostats, quadcopters, cameras, environmental sensors, robots, and all kinds of material goods both in the manufacturing process and in the distribution and retail system. This provides AI systems with far greater sensory and control access to the real world.

Finally, improvements in perception have allowed AI-powered robots to move out of the factory, where they relied on rigidly constrained arrangements of objects, and into the real, unstructured, messy world, where their cameras have something interesting to look at.

Self-driving cars

In the late 1950s, John McCarthy imagined that an automated vehicle might one day take him to the airport. In 1987, Ernst Dickmanns demonstrated a self-driving Mercedes van on the autobahn in Germany; it was capable of staying in lane, following another car, changing lanes, and overtaking.[3] More than thirty years later, we still don't have a fully autonomous car, but it's getting much closer. The focus of development has long since moved from academic research labs to large corporations. As of 2019, the best-performing test vehicles have logged millions of miles of driving on public roads (and billions of miles in driving simulators) without serious incident.[4] Unfortunately, other autonomous and semi-autonomous vehicles have killed several people.[5]

Why has it taken so long to achieve safe autonomous driving? The first reason is that the performance requirements are exacting.

Human drivers in the United States suffer roughly one fatal accident per one hundred million miles traveled, which sets a high bar. Autonomous vehicles, to be accepted, will need to be much better than that: perhaps one fatal accident per billion miles, or twenty-five thousand years of driving forty hours per week. The second reason is that one anticipated workaround—handing control to the human when the vehicle is confused or out of its safe operating conditions—simply doesn't work. When the car is driving itself, humans quickly become disengaged from the immediate driving circumstances and cannot regain context quickly enough to take over safely. Moreover, nondrivers and taxi passengers who are in the back seat are in no position to drive the car if something goes wrong.

Current projects are aiming at SAE Level 4 autonomy,[6] which means that the vehicle must at all times be capable of driving autonomously or stopping safely, subject to geographical limits and weather conditions. Because weather and traffic conditions can change, and because unusual circumstances can arise that a Level 4 vehicle cannot handle, a human has to be in the vehicle and ready to take over if needed. (Level 5—unrestricted autonomy—does not require a human driver but is even more difficult to achieve.) Level 4 autonomy goes far beyond the simple, reflex tasks of following white lines and avoiding obstacles. The vehicle has to assess the intent and probable future trajectories of all relevant objects, including objects that may not be visible, based on both current and past observations. Then, using look-ahead search, the vehicle has to find a trajectory that optimizes some combination of safety and progress. Some projects are trying more direct approaches based on reinforcement learning (mainly in simulation, of course) and supervised learning from recordings of hundreds of human drivers, but these approaches seem unlikely to reach the required level of safety.

The potential benefits of fully autonomous vehicles are immense. Every year, 1.2 million people die in car accidents worldwide and tens of millions suffer serious injuries. A reasonable target for autonomous vehicles would be to reduce these numbers by a factor of ten. Some analyses also predict a vast reduction in transportation costs, parking structures, congestion, and pollution. Cities will shift from personal cars and large buses to ubiquitous shared-ride, autonomous electric vehicles, providing door-to-door service and feeding high-speed mass-transit connections between hubs.[7] With costs as low as three cents per passenger mile, most cities would probably opt to provide the service for free—while subjecting riders to interminable barrages of advertising.

Of course, to reap all these benefits, the industry has to pay attention to the risks. If there are too many deaths attributed to poorly designed experimental vehicles, regulators may halt planned deployments or impose extremely stringent standards that might be unreachable for decades.[8] And people might, of course, decide not to buy or ride in autonomous vehicles unless they are demonstrably safe. A 2018 poll revealed a significant decline in consumers' level of trust in autonomous vehicle technology compared to 2016.[9] Even if the technology is successful, the transition to widespread autonomy will be an awkward one: human driving skills may atrophy or disappear, and the reckless and antisocial act of driving a car oneself may be banned altogether.

Intelligent personal assistants

Most readers will by now have experienced the unintelligent personal assistant: the smart speaker that obeys purchase commands overheard on the television, or the cell phone chatbot that responds to "Call me an ambulance!" with "OK, *from now on I'll call you 'Ann Ambulance.'*" Such systems are essentially voice-mediated interfaces to applications and search engines; they are based largely on canned stimulus–response templates, an approach that dates back to the Eliza system in the mid-1960s.[10]

These early systems have shortcomings of three kinds: access, content, and context. *Access shortcomings* mean that they lack sensory awareness of what's going on—for example, they might be able to hear what the user is saying but they can't see who the user is talking to. *Content shortcomings* mean that they simply fail to understand the meaning of what the user is saying or texting, even if they have access to it. *Context shortcomings* mean that they lack the ability to keep track of and reason about the goals, activities, and relationships that constitute daily life.

Despite these shortcomings, smart speakers and cell phone assistants offer just enough value to the user to have entered the homes and pockets of hundreds of millions of people. They are, in a sense, Trojan horses for AI. Because they are there, embedded in so many lives, every tiny improvement in their capabilities is worth billions of dollars.

And so, improvements are coming thick and fast. Probably the most important is the elementary capacity to understand content—to know that "John's in the hospital" is not just a prompt to say "*I hope it's nothing serious*" but contains actual information that the user's eight-year-old son is in a nearby hospital and may have a serious injury or illness. The ability to access email and text communications as well as phone calls and domestic conversations (through the smart speaker in the house) would give AI systems enough information to build a reasonably complete picture of the user's life—perhaps even more information than might have been available to the butler working for a nineteenth-century aristocratic family or the executive assistant working for a modern-day CEO.

Raw information, of course, is not enough. To be really useful, an assistant also needs commonsense knowledge of how the world works: that a child in the hospital is not simultaneously at home; that hospital care for a broken arm seldom lasts for more than a day or two; that the child's school will need to know of the expected absence; and so on. Such knowledge allows the assistant to keep track of things it does not observe directly—an essential skill for intelligent systems.

The capabilities described in the preceding paragraph are, I believe, feasible with existing technology for probabilistic reasoning,[C] but this would require a very substantial effort to construct models of all the kinds of events and transactions that make up our daily lives. Up to now, these kinds of commonsense modeling projects have generally not been undertaken (except possibly in classified systems for intelligence analysis and military planning) because of the costs involved and the uncertain payoff. Now, however, projects like this could easily reach hundreds of millions of users, so the investment risks are lower and the potential rewards are much higher. Furthermore, access to large numbers of users allows the intelligent assistant to learn very quickly and fill in all the gaps in its knowledge.

Thus, one can expect to see intelligent assistants that will, for pennies a month, help users with managing an increasingly large range of daily activities: calendars, travel, household purchases, bill payment, children's homework, email and call screening, reminders, meal planning, and—one can but dream—finding my keys. These skills will not be scattered across multiple apps. Instead, they will be facets of a single, integrated agent that can take advantage of the synergies available in what military people call the *common operational picture*.

The general design template for an intelligent assistant involves background knowledge about human activities, the ability to extract information from streams of perceptual and textual data, and a learning process to adapt the assistant to the user's particular circumstances. The same general template can be applied to at least three other major areas: health, education, and finances. For these applications, the system needs to keep track of the state of the user's body, mind, and bank account (broadly construed). As with assistants for daily life, the up-front cost of creating the necessary general knowledge in each of these three areas amortizes across billions of users.

In the case of health, for example, we all have roughly the same physiology, and detailed knowledge of how it works has already been encoded in machine-readable form.[11] Systems will adapt to your

individual characteristics and lifestyle, providing preventive suggestions and early warning of problems.

In the area of education, the promise of intelligent tutoring systems was recognized even in the 1960s,[12] but real progress has been a
long time coming. The primary reasons are shortcomings of content
and access: most tutoring systems don't understand the content of
what they purport to teach, nor can they engage in two-way communication with their pupils through speech or text. (I imagine myself
teaching string theory, which I don't understand, in Laotian, which I
don't speak.) Recent progress in speech recognition means that automated tutors can, at last, communicate with pupils who are not yet
fully literate. Moreover, probabilistic reasoning technology can now
keep track of what students know and don't know[13] and can optimize
the delivery of instruction to maximize learning. The Global Learning
XPRIZE competition, which started in 2014, offered $15 million for
"open-source, scalable software that will enable children in developing countries to teach themselves basic reading, writing and arithmetic within 15 months." Results from the winners, Kitkit School and
onebillion, suggest that the goal has largely been achieved.

In the area of personal finance, systems will keep track of investments, income streams, obligatory and discretionary expenditures,
debt, interest payments, emergency reserves, and so on, in much the
same way that financial analysts keep track of the finances and prospects of corporations. Integration with the agent that handles daily life
will provide an even finer-grained understanding, perhaps even ensuring that the children get their pocket money minus any mischief-
related deductions. One can expect to receive the quality of day-to-day
financial advice previously reserved for the ultra-rich.

If your privacy alarm bells weren't ringing as you read the preceding paragraphs, you haven't been keeping up with the news. There are,
however, multiple layers to the privacy story. First, can a personal
assistant really be useful if it knows nothing about you? Probably not.

Second, can personal assistants be really useful if they cannot pool information from multiple users to learn more about people in general and people who are similar to you? Probably not. So, don't those two things imply that we have to give up our privacy to benefit from AI in our daily lives? No. The reason is that learning algorithms can operate on *encrypted* data using the techniques of secure multiparty computation, so that users can benefit from pooling without compromising privacy in any way.[14] Will software providers adopt privacy-preserving technology voluntarily, without legislative encouragement? That remains to be seen. What seems inevitable, however, is that users will trust a personal assistant only if its primary obligation is to the user rather than to the corporation that produced it.

Smart homes and domestic robots

The smart home concept has been investigated for several decades. In 1966, James Sutherland, an engineer at Westinghouse, started collecting surplus computer parts to build ECHO, the first smart-home controller.[15] Unfortunately, ECHO weighed eight hundred pounds, consumed 3.5 kilowatts, and managed just three digital clocks and the TV antenna. Subsequent systems required users to master control interfaces of mind-boggling complexity. Unsurprisingly, they never caught on.

Beginning in the 1990s, several ambitious projects attempted to design houses that managed themselves with minimal human intervention, using machine learning to adapt to the lifestyles of the occupants. To make these experiments meaningful, real people had to live in the houses. Unfortunately, the frequency of erroneous decisions made the systems worse than useless—the occupants' quality of life decreased rather than increased. For example, inhabitants of the 2003 MavHome project[16] at Washington State University often had to sit in the dark if their visitors stayed later than the usual bedtime.[17] As with the unintelligent personal assistant, such failings result from inadequate sensory access to the activities of the occupants and the inability to understand and keep track of what's happening in the house.

A truly smart home equipped with cameras and microphones—and the requisite perceptual and reasoning abilities—can understand what the occupants are doing: visiting, eating, sleeping, watching TV, reading, exercising, getting ready for a long trip, or lying helpless on the floor after a fall. By coordinating with the intelligent personal assistant, the home can have a pretty good idea of who will be in or out of the house at what time, who's eating where, and so on. This understanding allows it to manage heating, lighting, window blinds, and security systems, to send timely reminders, and to alert users or emergency services when a problem arises. Some newly built apartment complexes in the United States and Japan are already incorporating technology of this kind.[18]

The value of the smart home is limited because of its actuators: much simpler systems (timed thermostats and motion-sensitive lights and burglar alarms) can deliver a lot of the same functionality in ways that are perhaps more predictable, if less context sensitive. The smart home cannot fold the laundry, clear the dishes, or pick up the newspaper. It *really* wants a physical robot to do its bidding.

 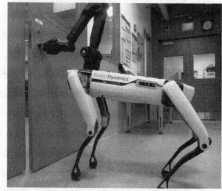

FIGURE 5: (left) BRETT folding towels; (right) the Boston Dynamics SpotMini robot opening a door.

It may not have too long to wait. Already, robots have demonstrated many of the required skills. In the Berkeley lab of my colleague Pieter Abbeel, BRETT (the Berkeley Robot for the Elimination of Tedious Tasks) has been folding piles of towels since 2011, while the SpotMini robot from Boston Dynamics can climb stairs and open doors (figure 5). Several companies are already building cooking robots, although they require special, enclosed setups and pre-cut ingredients and won't work in an ordinary kitchen.[19]

Of the three basic physical capabilities required for a useful domestic robot—perception, mobility, and dexterity—the latter is most problematic. As Stefanie Tellex, a robotics professor at Brown University, puts it, "Most robots can't pick up most objects most of the time." This is partly a problem of tactile sensing, partly a manufacturing problem (dexterous hands are currently very expensive to build), and partly an algorithmic problem: we don't yet have a good understanding of how to combine sensing and control to grasp and manipulate the huge variety of objects in a typical household. There are dozens of grasp types just for rigid objects and there are thousands of distinct manipulation skills, such as shaking exactly two pills out of a bottle, peeling the label off a jam jar, spreading hard butter on soft bread, or lifting one strand of spaghetti from the pot with a fork to see if it's ready.

It seems likely that the tactile sensing and hand construction problems will be solved by 3D printing, which is already being used by Boston Dynamics for some of the more complex parts of their Atlas humanoid robot. Robot manipulation skills are advancing rapidly, thanks in part to deep reinforcement learning.[20] The final push—putting all this together into something that begins to approximate the awesome physical skills of movie robots—is likely to come from the rather unromantic warehouse industry. Just one company, Amazon, employs several hundred thousand people who pick products out of bins in giant warehouses and dispatch them to customers. From 2015 through 2017 Amazon ran an annual "Picking Challenge" to accelerate the development of robots capable of doing this task.[21] There is still some distance to go, but when the core research problems are solved—probably within a decade—one can expect a very rapid rollout of highly capable robots. Initially they will work in warehouses, then in other commercial applications such as agriculture and construction, where the range of tasks and objects is fairly predictable. We might also see them quite soon in the retail sector doing tasks such as stocking supermarket shelves and refolding clothes.

The first to really benefit from robots in the home will be the elderly and infirm, for whom a helpful robot can provide a degree of independence that would otherwise be impossible. Even if the robot has a limited repertoire of tasks and only rudimentary comprehension of what's going on, it can still be very useful. On the other hand, the robot butler, managing the household with aplomb and anticipating its master's every wish, is still some way off—it requires something approaching the generality of human-level AI.

Intelligence on a global scale

The development of basic capabilities for understanding speech and text will allow intelligent personal assistants to do things that human assistants can already do (but they will be doing it for pennies per month instead of thousands of dollars per month). Basic speech and text understanding also enable machines to do things that no human can do—not because of the *depth* of understanding but because of its *scale*. For example, a machine with basic reading capabilities will be able to read *everything the human race has ever written* by lunchtime, and then it will be looking around for something else to do.[22] With speech recognition capabilities, it could listen to *every radio and television broadcast* before teatime. For comparison, it would take two hundred thousand full-time humans just to keep up with the world's current level of print publication (let alone all the written material from the past) and another sixty thousand to listen to current broadcasts.[23]

Such a system, if it could extract even simple factual assertions and integrate all this information across all languages, would represent an incredible resource for answering questions and revealing patterns— probably far more powerful than search engines, which are currently valued at around $1 trillion. Its research value for fields such as history and sociology would be inestimable.

Of course, it would also be possible to listen to all the world's phone calls (a job that would require about twenty million people). There are certain clandestine agencies that would find this valuable. Some of them have been doing simple kinds of large-scale machine listening, such as spotting key words in conversations, for many years, and have now made the transition to transcribing entire conversations into searchable text.[24] Transcriptions are certainly useful, but not nearly as useful as simultaneous understanding and content integration of *all* conversations.

Another "superpower" that is available to machines is to *see the entire world at once.* Roughly speaking, satellites image the entire world every day at an average resolution of around fifty centimeters per pixel. At this resolution, every house, ship, car, cow, and tree on Earth is visible. Well over thirty million full-time employees would be needed to examine all these images;[25] so, at present, no human ever sees the vast majority of satellite data. Computer vision algorithms could process all this data to produce a searchable database of the whole world, updated daily, as well as visualizations and predictive models of economic activities, changes in vegetation, migrations of animals and people, the effects of climate change, and so on. Satellite companies such as Planet and DigitalGlobe are busy making this idea a reality.

With the possibility of sensing on a global scale comes the possibility of decision making on a global scale. For example, from global satellite data feeds, it should be possible to create detailed models for managing the global environment, predicting the effects of environmental and economic interventions, and providing the necessary analytical inputs to the UN's sustainable development goals.[26] We are already seeing "smart city" control systems that aim to optimize traffic management, transit, trash collection, road repairs, environmental maintenance, and other functions for the benefit of citizens, and these may be extended to the country level. Until recently, this degree of coordination could be achieved only by huge, inefficient, bureaucratic hierarchies of humans; inevitably, these will be replaced by mega-agents that take care of more and more aspects of our collective lives. Along with this, of course, comes the possibility of privacy invasion and social control on a global scale, to which I return in the next chapter.

When Will Superintelligent AI Arrive?

I am often asked to predict when superintelligent AI will arrive, and I usually refuse to answer. There are three reasons for this. First, there is a long history of such predictions going wrong.[27] For example, in 1960, the AI pioneer and Nobel Prize–winning economist Herbert Simon wrote, "Technologically . . . machines will be capable, within twenty years, of doing any work a man can do."[28] In 1967, Marvin Minsky, a co-organizer of the 1956 Dartmouth workshop that started the field of AI, wrote, "Within a generation, I am convinced, few compartments of intellect will remain outside the machine's realm—the problem of creating 'artificial intelligence' will be substantially solved."[29]

A second reason for declining to provide a date for superintelligent AI is that there is no clear threshold that will be crossed. Machines already exceed human capabilities in some areas. Those areas will broaden and deepen, and it is likely that there will be superhuman general knowledge systems, superhuman biomedical research systems, superhuman dexterous and agile robots, superhuman corporate planning systems, and so on well before we have a completely general superintelligent AI system. These "partially superintelligent" systems will, individually and collectively, begin to pose many of the same issues that a generally intelligent system would.

A third reason for not predicting the arrival of superintelligent AI is that it is inherently unpredictable. It requires "conceptual breakthroughs," as noted by John McCarthy in a 1977 interview.[30] McCarthy went on to say, "What you want is 1.7 Einsteins and 0.3 of the Manhattan Project, and you want the Einsteins first. I believe it'll take five to 500 years." In the next section I'll explain what some of the conceptual breakthroughs are likely to be. Just how unpredictable are they? Probably as unpredictable as Szilard's invention of the nuclear chain reaction a few hours after Rutherford's declaration that it was completely impossible.

Once, at a meeting of the World Economic Forum in 2015, I answered the question of when we might see superintelligent AI. The meeting was under Chatham House rules, which means that no remarks may be attributed to anyone present at the meeting. Even so, out of an excess of caution, I prefaced my answer with "Strictly off the record. . . ." I suggested that, barring intervening catastrophes, it would probably happen in the lifetime of my children—who were still quite young and would probably have much longer lives, thanks to advances in medical science, than many of those at the meeting. Less than two hours later, an article appeared in the *Daily Telegraph* citing Professor Russell's remarks, complete with images of rampaging Terminator robots. The headline was 'SOCIOPATHIC' ROBOTS COULD OVERRUN THE HUMAN RACE WITHIN A GENERATION.

My timeline of, say, eighty years is considerably more conservative than that of the typical AI researcher. Recent surveys[31] suggest that most active researchers expect human-level AI to arrive around the middle of this century. Our experience with nuclear physics suggests that it would be prudent to assume that progress could occur quite quickly and to prepare accordingly. If just one conceptual breakthrough were needed, analogous to Szilard's idea for a neutron-induced nuclear chain reaction, superintelligent AI in some form could arrive quite suddenly. The chances are that we would be unprepared: if we built superintelligent machines with any degree of autonomy, we would soon find ourselves unable to control them. I am, however, fairly confident that we have some breathing space because there are several major breakthroughs needed between here and superintelligence, not just one.

Conceptual Breakthroughs to Come

The problem of creating general-purpose, human-level AI is far from solved. Solving it is not a matter of spending money on more engineers, more data, and bigger computers. Some futurists produce charts that extrapolate the exponential growth of computing power into the future based on Moore's law, showing the dates when machines will become more powerful than insect brains, mouse brains, human brains, *all* human brains put together, and so on.[32] These charts are meaningless because, as I have already said, faster machines just give you the wrong answer more quickly. If one were to collect AI's leading experts into a single team with unlimited resources, with the goal of creating an integrated, human-level intelligent system by combining all our best ideas, the result would be failure. The system would break in the real world. It wouldn't understand what was going on; it wouldn't be able to predict the consequences of its actions; it wouldn't understand what people want in any given situation; and so it would do ridiculously stupid things.

By understanding *how* the system would break, AI researchers are able to identify the problems that have to be solved—the conceptual breakthroughs that are needed—in order to reach human-level AI. I will now describe some of these remaining problems. Once they are solved, there may be more, but not very many more.

Language and common sense

Intelligence without knowledge is like an engine without fuel. Humans acquire a vast amount of knowledge from other humans: it is passed down through generations in the form of language. Some of it is factual: Obama became president in 2009, the density of copper is 8.92 grams per cubic centimeter, the code of Ur-Nammu set out punishments for various crimes, and so on. A great deal of knowledge resides in the language itself—in the concepts that it makes available. *President, 2009, density, copper, gram, centimeter, crime,* and the rest all carry with them a vast amount of information, which represents the extracted essence of the processes of discovery and organization that led them to be in the language in the first place.

Take, for example, *copper,* which refers to some collection of atoms in the universe, and compare it to *arglebarglium,* which is my name for an equally large collection of entirely randomly selected atoms in the universe. There are many general, useful, and predictive laws one can discover about copper—about its density, conductivity, malleability, melting point, stellar origin, chemical compounds, practical uses, and so on; in comparison, there is essentially nothing that can be said about arglebarglium. An organism equipped with a language composed of words like arglebarglium would be unable to function, because it would never discover the regularities that would allow it to model and predict its universe.

A machine that *really* understands human language would be in a position to quickly acquire vast quantities of human knowledge, allowing it to bypass tens of thousands of years of learning by the more than one hundred billion people who have lived on Earth. It seems simply impractical to expect a machine to rediscover all this from scratch, starting from raw sensory data.

At present, however, natural language technology is not up to the task of reading and understanding millions of books—many of which would stump even a well-educated human. Systems such as IBM's Watson, which famously defeated two human champions of the *Jeopardy!* quiz game in 2011, can extract simple information from clearly stated facts but cannot build complex knowledge structures from text; nor can they answer questions that require extensive chains of reasoning with information from multiple sources. For example, the task of reading all available documents up to the end of 1973 and assessing (with explanations) the probable outcome of the Watergate impeachment process against then president Nixon would be well beyond the current state of the art.

There are serious efforts underway to deepen the level of language analysis and information extraction. For example, Project Aristo at the Allen Institute for AI aims to build systems that can pass school science exams after reading textbooks and study guides.[33] Here's a question from a fourth-grade test:[34]

Fourth graders are planning a roller-skate race. Which surface would be the best for this race?

(A) gravel   (B) sand   (C) blacktop   (D) grass

A machine faces at least two sources of difficulty in answering this question. The first is the classical language-understanding problem of working out what the sentences say: analyzing the syntactic structure, identifying the meanings of words, and so on. (Try this for yourself: use an online translation service to translate the sentences into an unfamiliar language, then use a dictionary for that language to try translating them back to English.) The second is the need for common-sense knowledge: to work out that a "roller-skate race" is probably a race between people wearing roller skates (on their feet) rather than a race between roller skates, to understand that the "surface" is what the skaters will skate on rather than what the spectators will sit on, to know what "best" means in the context of a surface for a race, and so on. Think how the answer might change if we replaced "fourth grad-ers" with "sadistic army boot-camp trainers."

One way to summarize the difficulty is to say that reading requires knowledge and knowledge (largely) comes from reading. In other words, we face a classic chicken-and-egg situation. We might hope for a bootstrapping process, whereby the system reads some easy text, acquires some knowledge, uses that to read more difficult text, ac-quires still more knowledge, and so on. Unfortunately, what tends to happen is the opposite: the knowledge acquired is mostly erroneous, which causes errors in reading, which results in more erroneous knowl-edge, and so on.

For example, the NELL (Never-Ending Language Learning) proj-ect at Carnegie Mellon University is probably the most ambitious language-bootstrapping project currently underway. From 2010 to 2018, NELL acquired over 120 million beliefs by reading English text on the Web.[35] Some of these beliefs are accurate, such as the beliefs that the Maple Leafs play hockey and won the Stanley Cup. In addi-tion to facts, NELL acquires new vocabulary, categories, and semantic relationships all the time. Unfortunately, NELL has confidence in only 3 percent of its beliefs and relies on human experts to clean out false or meaningless beliefs on a regular basis—such as its beliefs that "*Nepal* is a *country* also known as *United States*" and "*value* is an *agricultural product* that is usually cut into *basis*."

I suspect that there may be no single breakthrough that turns the downward spiral into an upward spiral. The basic bootstrapping pro-cess seems right: a program that knows enough facts can figure out which fact a novel sentence is referring to, and thereby learns a new textual form for expressing facts—which then lets it discover more facts, and so the process continues. (Sergey Brin, the co-founder of Google, published an important paper on the bootstrapping idea in 1998.[36]) Priming the pump by supplying a good deal of manually en-coded knowledge and linguistic information would certainly help.

Increasing the sophistication of the representation of facts—allowing for complex events, causal relationships, beliefs and attitudes of others, and so on—and improving the handling of uncertainty about word meanings and sentence meanings may eventually result in a self-reinforcing rather than self-extinguishing process of learning.

Cumulative learning of concepts and theories

Approximately 1.4 billion years ago and 8.2 sextillion miles away, two black holes, one twelve million times the mass of the Earth and the other ten million, came close enough to begin orbiting each other. Gradually losing energy, they spiraled closer and closer to each other and faster and faster, reaching an orbital frequency of 250 times per second at a distance of 350 kilometers before finally colliding and merging.[37] In the last few milliseconds, the rate of energy emission in the form of gravitational waves was fifty times larger than the total energy output of all the stars in the universe. On September 14, 2015, those gravitational waves arrived at the Earth. They alternately expanded and compressed space itself by a factor of about one in 2.5 sextillion, equivalent to changing the distance to Proxima Centauri (4.4 light years) by the width of a human hair.

Fortunately, two days earlier, the Advanced LIGO (Laser Interferometer Gravitational-Wave Observatory) detectors in Washington and Louisiana had been switched on. Using laser interferometry, they were able to measure the minuscule distortion of space; using calculations based on Einstein's theory of general relativity, the LIGO researchers had predicted—and were therefore looking for—the exact shape of the gravitational waveform expected from such an event.[38]

This was possible because of the accumulation and communication of knowledge and concepts by thousands of people across centuries of observation and research. From Thales of Miletus rubbing amber with wool and observing the static charge buildup, through

Galileo dropping rocks from the Leaning Tower of Pisa, to Newton seeing an apple fall from a tree, and on through thousands more observations, humanity has gradually accumulated layer upon layer of concepts, theories, and devices: mass, velocity, acceleration, force, Newton's laws of motion and gravitation, orbital equations, electrical phenomena, atoms, electrons, electric fields, magnetic fields, electromagnetic waves, special relativity, general relativity, quantum mechanics, semiconductors, lasers, computers, and so on.

Now, *in principle* we can understand this process of discovery as a mapping from all the sensory data ever experienced by all humans to a very complex hypothesis about the sensory data experienced by the LIGO scientists on September 14, 2015, as they watched their computer screens. This is the purely data-driven view of learning: data in, hypothesis out, black box in between. If it could be done, it would be the apotheosis of the "big data, big network" deep learning approach, but it cannot be done. The only plausible idea we have for how intelligent entities could achieve such a stupendous feat as detecting the merger of two black holes is that *prior knowledge of physics*, combined with the observational data from their instruments, allowed the LIGO scientists to infer the occurrence of the merger event. Moreover, this prior knowledge was itself the result of learning with prior knowledge— and so on, all the way back through history. Thus, we have a roughly *cumulative* picture of how intelligent entities can build predictive capabilities, with knowledge as the building material.

I say *roughly* because, of course, science has taken a few wrong turns over the centuries, temporarily pursuing illusory notions such as phlogiston and the luminiferous aether. But we know for a fact that the cumulative picture is what *actually* happened, in the sense that scientists all along the way wrote down their findings and theories in books and papers. Later scientists had access only to these forms of explicit knowledge, and not to the original sensory experiences of earlier, long-dead generations. Because they are scientists, the members of the LIGO team understood that all the pieces of knowledge they used, including Einstein's theory of general relativity, are (and always will be) in their probationary period and *could* be falsified by experiment. As it turned out, the LIGO data provided strong confirmation for general relativity as well as further evidence that the graviton—a hypothesized particle that mediates the force of gravity—is massless.

We are a very long way from being able to create machine learning systems that are capable of matching or exceeding the capacity for cumulative learning and discovery exhibited by the scientific community—or by ordinary human beings in their own lifetimes.[39] Deep learning systems[D] are mostly data driven: at best, we can "wire in" some very weak forms of prior knowledge in the structure of the network. Probabilistic programming systems[C] do allow for prior knowledge in the learning process, as expressed in the structure and vocabulary of the probabilistic knowledge base, but we do not yet have effective methods for generating new concepts and relationships and using them to expand such a knowledge base.

The difficulty is not one of finding hypotheses that provide a good fit to data; deep learning systems can find hypotheses that are a good fit to image data, and AI researchers have built symbolic learning programs able to recapitulate many historical discoveries of quantitative scientific laws.[40] Learning in an autonomous intelligent agent requires much more than this.

First, what should be included in the "data" from which predictions are made? For example, in the LIGO experiment, the model for predicting the amount that space stretches and shrinks when a gravitational wave arrives takes into account the masses of the colliding black holes, the frequency of their orbits, and so on, but it doesn't take into account the day of the week or the occurrence of Major League baseball games. On the other hand, a model for predicting traffic on the San Francisco Bay Bridge takes into account the day of the week and the occurrence of Major League baseball games but ignores the masses and orbital frequencies of colliding black holes. Similarly, programs that learn to recognize the *types* of objects in images use the pixels as input, whereas a program that learns to estimate the *value* of an antique object would also want to know what it was made of, who made it and when, its history of usage and ownership, and so on. Why is this? Obviously, it's because we humans already know something about gravitational waves, traffic, visual images, and antiques. We use this knowledge to decide which inputs are needed for predicting a specific output. This is called *feature engineering*, and doing it well requires a good understanding of the specific prediction problem.

Of course, a real intelligent machine cannot rely on human feature engineers showing up every time there is something new to learn. It will have to work out for itself what constitutes a reasonable hypothesis space for a learning problem. Presumably, it will do this by bringing to bear a wide range of relevant knowledge in various forms, but at present we have only rudimentary ideas about how to do this.[41] Nelson Goodman's *Fact, Fiction, and Forecast*[42]—written in 1954 and perhaps one of the most important and underappreciated books on machine learning—suggests a kind of knowledge called an *overhypothesis*, because it helps to define what the space of reasonable hypotheses might be. In the case of traffic prediction, for example, the relevant overhypothesis would be that the day of the week, time of day, local events, recent accidents, holidays, transit delays, weather, and sunrise and sunset times can influence traffic conditions. (Notice that you can figure out this overhypothesis from your own background knowledge of the world, without being a traffic expert.) An intelligent learning system can accumulate and use knowledge of this kind to help formulate and solve new learning problems.

Second, and perhaps more important, is the cumulative generation of new concepts such as mass, acceleration, charge, electron, and gravitational force. Without these concepts, scientists (and ordinary people) would have to interpret their universe and make predictions on the basis of raw perceptual inputs. Instead, Newton was able to work with concepts of mass and acceleration developed by Galileo and others; Rutherford could determine that the atom was composed of a dense, positively charged nucleus surrounded by electrons because the concept of an electron had already been developed (by numerous researchers in small steps) in the late nineteenth century; indeed, all scientific discoveries rely on layer upon layer of concepts that stretch back through time and human experience.

In the philosophy of science, particularly in the early twentieth century, it was not uncommon to see the discovery of new concepts attributed to the three ineffable I's: intuition, insight, and inspiration. All these were considered resistant to any rational or algorithmic explanation. AI researchers, including Herbert Simon,[43] have objected strongly to this view. Put simply, if a machine learning algorithm can search in a space of hypotheses that includes the possibility of adding definitions for new terms not present in the input, then the algorithm can discover new concepts.

For example, suppose that a robot is trying to learn the rules of backgammon by watching people playing the game. It observes how they roll the dice and notices that sometimes players move three or four pieces rather than one or two and that this happens after a roll of 1-1, 2-2, 3-3, 4-4, 5-5, or 6-6. If the program can add a new concept of *doubles*, defined by equality between the two dice, it can express the same predictive theory much more concisely. It is a straightforward process, using methods such as inductive logic programming,[44] to create programs that propose new concepts and definitions in order to identify theories that are both accurate and concise.

At present, we know how to do this for relatively simple cases, but for more complex theories the number of possible new concepts that could be introduced becomes simply enormous. This makes the recent success of deep learning methods in computer vision all the more intriguing. The deep networks usually succeed in finding useful intermediate features such as eyes, legs, stripes, and corners, even though they are using very simple learning algorithms. If we can understand better how this happens, we can apply the same approach to learning new concepts in the more expressive languages needed for science. This by itself would be a huge boon to humanity as well as a significant step towards general-purpose AI.

Discovering actions

Intelligent behavior over long time scales requires the ability to plan and manage activity hierarchically, at multiple levels of abstraction—all the way from doing a PhD (one trillion actions) to a single motor control command sent to one finger as part of typing a single character in the application cover letter.

Our activities are organized into complex hierarchies with *dozens* of levels of abstraction. These levels and the actions they contain are a key part of our civilization and are handed down through generations via our language and practices. For example, actions such as *catching a wild boar* and *applying for a visa* and *buying a plane ticket* may involve millions of primitive actions, but we can think about them as single units because they are already in the "library" of actions that our language and culture provides and because we know (roughly) how to do them.

Once they are in the library, we can string these high-level actions together into still higher-level actions, such as having a tribal feast for the summer solstice or doing archaeological research for a summer in a remote part of Nepal. Trying to plan such activities from scratch, starting with the lowest-level motor control steps, would be completely hopeless because such activities involve millions or billions of steps, many of which are very unpredictable. (Where will the wild boar be found, and which way will he run?) With suitable high-level actions in the library, on the other hand, one need plan only a dozen or so steps, because each such step is a large piece of the overall activity. This is something that even our feeble human brains can manage—but it gives us the "superpower" of planning over long time scales.

There was a time when these actions didn't exist as such—for

FIGURE 6: Saul Steinberg's *View of the World from 9th Avenue*, 1976, first published as a cover of *The New Yorker* magazine.

example, to obtain the right to a plane journey in 1910 would have required a long, involved, and unpredictable process of research, letter writing, and negotiation with various aeronautical pioneers. Other actions recently added to the library include emailing, googling, and ubering. As Alfred North Whitehead wrote in 1911, "Civilization advances by extending the number of important operations which we can perform without thinking about them."[45]

Saul Steinberg's famous cover for *The New Yorker* (figure 6) brilliantly shows, in spatial form, how an intelligent agent manages its own future. The very immediate future is extraordinarily detailed—in fact, my brain has already loaded up the specific motor control sequences for typing the next few words. Looking a bit further ahead, there is less detail—my plan is to finish this section, have lunch, write some more, and watch France play Croatia in the final of the World Cup. Still further ahead, my plans are larger but vaguer: move back from Paris to Berkeley in early August, teach a graduate course, and finish this book. As one moves through time, the future moves closer to the present and the plans for it become more detailed, while new, vague plans may be added to the distant future. Plans for the immediate future become so detailed that they are executable directly by the motor control system.

At present we have only some pieces of this overall picture in place for AI systems. If the hierarchy of abstract actions is provided—including knowledge of how each abstract action can be refined into a subplan composed of more concrete actions—then we have algorithms that can construct complex plans to achieve specific goals. There are algorithms that can execute abstract, hierarchical plans in such a way that the agent always has a primitive, physical action "ready to go," even if actions in the future are still at an abstract level and not yet executable.

The main missing piece of the puzzle is a method for constructing the hierarchy of abstract actions in the first place. For example, is it possible to start from scratch with a robot that knows only that it can send various electric currents to various motors and have it discover for itself the action of standing up? It's important to understand that I'm *not* asking whether we can train a robot to stand up, which can be done simply by applying reinforcement learning with a reward for the robot's head being farther away from the ground.[46] Training a robot to stand up requires that the human trainer already knows what *standing up* means, so that the right reward signal can be defined. What we want is for the robot to discover for itself that *standing up* is a thing—a useful abstract action, one that achieves the precondition (being upright) for walking or running or shaking hands or seeing over a wall and so forms part of many abstract plans for all kinds of goals.

Similarly, we want the robot to discover actions such as moving from place to place, picking up objects, opening doors, tying knots, cooking dinner, finding my keys, building houses, and many other actions that have no names in any human language because we humans have not discovered them yet.

I believe this capability is the most important step needed to reach human-level AI. It would, to borrow Whitehead's phrase again, extend the number of important operations that AI systems can perform without thinking about them. Numerous research groups around the world are hard at work on solving the problem. For example, Deep-Mind's 2018 paper showing human-level performance on Quake III Arena Capture the Flag claims that their learning system "constructs a temporally hierarchical representation space in a novel way to promote . . . temporally coherent action sequences."[47] (I'm not completely sure what this means, but it certainly sounds like progress towards the goal of inventing new high-level actions.) I suspect that we do not yet have the complete answer, but this is an advance that could occur any moment, just by putting some existing ideas together in the right way.

Intelligent machines with this capability would be able to look further into the future than humans can. They would also be able to take into account far more information. These two capabilities combined lead inevitably to better real-world decisions. In any kind of conflict situation between humans and machines, we would quickly find, like Garry Kasparov and Lee Sedol, that our every move has been anticipated and blocked. We would lose the game before it even started.

Managing mental activity

If managing activity in the real world seems complex, spare a thought for your poor brain, managing the activity of the "most complex object in the known universe"—itself. We don't start out knowing how to think, any more than we start out knowing how to walk or play the piano. We learn how to do it. We can, to some extent, *choose* what thoughts to have. (Go on, think about a juicy hamburger or Bulgarian customs regulations—your choice!) In some ways, our mental activity is more complex than our activity in the real world, because our brains have far more moving parts than our bodies and those parts move much faster. The same is true for computers: for every move that AlphaGo makes on the Go board, it performs *millions* or *billions* of units of computation, each of which involves adding a branch to the lookahead search tree and evaluating the board position at the end of that branch. And each of those units of computation happens because the program makes a *choice* about which part of the tree to explore next. Very approximately, AlphaGo chooses computations that it expects will improve its eventual decision on the board.

It has been possible to work out a reasonable scheme for managing AlphaGo's computational activity because that activity is simple and homogeneous: every unit of computation is of the same kind. Compared to other programs that use that same basic unit of computation, AlphaGo is probably quite efficient, but it's probably extremely *inefficient* compared to other kinds of programs. For example, Lee Sedol, AlphaGo's human opponent in the epochal match of 2016, probably does no more than a few thousand units of computation per move, but he has a much more flexible computational architecture with many more kinds of units of computation: these include dividing the board into subgames and trying to resolve their interactions; recognizing possible goals to attain and making high-level plans with actions like "keep this group alive" or "prevent my opponent from connecting these two groups"; thinking about *how* to achieve a specific goal, such as keeping a group alive; and ruling out whole classes of moves because they fail to address a significant threat.

We simply don't know how to organize such complex and varied computational activity—how to integrate and build on the results from each and how to allocate computational resources to the various kinds of deliberation so that good decisions are found as quickly as possible. It is clear, however, that a simple computational architecture like AlphaGo's cannot possibly work in the real world, where we routinely need to deal with decision horizons of not tens but billions of primitive steps and where the number of possible actions at any point is almost infinite. It's important to remember that an intelligent agent in the real world is not restricted to *playing Go* or even *finding Stuart's keys*—it's just *being*. It can do *anything* next, but it cannot possibly afford to think about all the things it might do.

A system that can both discover new high-level actions—as described earlier—and manage its computational activity to focus on units of computation that quickly deliver significant improvements in decision quality would be a formidable decision maker in the real world. Like those of humans, its deliberations would be "cognitively efficient," but it would not suffer from the tiny short-term memory and slow hardware that severely limit our ability to look far into the future, handle a large number of contingencies, and consider a large number of alternative plans.

More things missing?

If we put together everything we know how to do with all the potential new developments listed in this chapter, would it work? How would the resulting system behave? It would plow through time, absorbing vast quantities of information and keeping track of the state of the world on a massive scale by observation and inference. It would gradually improve its models of the world (which include models of humans, of course). It would use those models to solve complex problems and it would encapsulate and reuse its solution processes to make its deliberations more efficient and to enable the solution of still more complex problems. It would discover new concepts and actions, and these would allow it to improve its rate of discovery. It would make effective plans over increasingly long time scales.

In summary, it's not obvious that anything else of great significance is missing, from the point of view of systems that are effective in achieving their objectives. Of course, the only way to be sure is to build it (once the breakthroughs have been achieved) and see what happens.

Imagining a Superintelligent Machine

The technical community has suffered from a failure of imagination when discussing the nature and impact of superintelligent AI. Often, we see discussions of reduced medical errors,[48] safer cars,[49] or other advances of an incremental nature. Robots are imagined as individual entities carrying their brains with them, whereas in fact they are likely to be wirelessly connected into a single, global entity that draws on vast stationary computing resources. It's as if researchers are afraid of examining the real consequences of success in AI.

A general-purpose intelligent system can, by assumption, do what any human can do. For example, some humans did a lot of mathematics, algorithm design, coding, and empirical research to come up with the modern search engine. The results of all this work are very useful and of course very valuable. How valuable? A recent study showed that the median American adult surveyed would need to be paid at least $17,500 to give up using search engines for a year,[50] which translates to a global value in the tens of trillions of dollars.

Now imagine that search engines don't exist yet because the necessary decades of work have not been done, but you have access instead to a superintelligent AI system. Simply by asking the question, you now have access to search engine technology, courtesy of the AI system. Done! Trillions of dollars in value, just for the asking, and not a single line of additional code written by you. The same goes for any other missing invention or series of inventions: if humans could do it, so can the machine.

This last point provides a useful lower bound—a pessimistic

estimate—on what a superintelligent machine can do. By assumption, the machine is more capable than an individual human. There are many things an individual human cannot do, but a collection of n humans can do: put an astronaut on the Moon, create a gravitational-wave detector, sequence the human genome, run a country with hundreds of millions of people. So, roughly speaking, we create n software copies of the machine and connect them in the same way—with the same information and control flows—as the n humans. Now we have a machine that can do whatever n humans can do, except better, because each of its n components is superhuman.

This *multi-agent cooperation* design for an intelligent system is just a lower bound on the possible capabilities of machines because there are other designs that work better. In a collection of n humans, the total available information is kept separately in n brains and communicated very slowly and imperfectly between them. That's why the n humans spend most of their time in meetings. In the machine, there is no need for this separation, which often prevents connecting the dots. For an example of disconnected dots in scientific discovery, a brief perusal of the long history of penicillin is quite eye-opening.[51]

Another useful method of stretching your imagination is to think about some particular form of sensory input—say, reading—and scale it up. Whereas a human can read and understand one book in a week, a machine could read and understand every book ever written—all 150 million of them—in a few hours. This requires a decent amount of processing power, but the books can be read largely in parallel, meaning that simply adding more chips allows the machine to scale up its reading process. By the same token, the machine can see everything at once through satellites, robots, and hundreds of millions of surveillance cameras; watch all the world's TV broadcasts; and listen to all the world's radio stations and phone conversations. Very quickly it would gain a far more detailed and accurate understanding of the world and its inhabitants than any human could possibly hope to acquire.

One can also imagine scaling the machine's capacity for action. A human has direct control over only one body, while a machine can control thousands or millions. Some automated factories already exhibit this characteristic. Outside the factory, a machine that controls thousands of dexterous robots can, for example, produce vast numbers of houses, each one tailored to its future occupants' needs and desires. In the lab, existing robotic systems for scientific research could be scaled up to perform millions of experiments simultaneously—perhaps to create complete predictive models of human biology down to the molecular level. Note that the machine's reasoning capabilities will give it a far greater capacity to detect inconsistencies between scientific theories and between theories and observations. Indeed, it may already be the case that we have enough experimental evidence about biology to devise a cure for cancer: we just haven't put it together.

In the cyber realm, machines already have access to billions of effectors—namely, the displays on all the phones and computers in the world. This partly explains the ability of IT companies to generate enormous wealth with very few employees; it also points to the severe vulnerability of the human race to manipulation via screens.

Scale of a different kind comes from the machine's ability to look further into the future, with greater accuracy, than is possible for humans. We have seen this for chess and Go already; with the capacity for generating and analyzing hierarchical plans over long time scales and the ability to identify new abstract actions and high-level descriptive models, machines will transfer this advantage to domains such as mathematics (proving novel, useful theorems) and decision making in the real world. Tasks such as evacuating a large city in the event of an environmental disaster will be relatively straightforward, with the machine able to generate individual guidance for every person and vehicle to minimize the number of casualties.

The machine might work up a slight sweat when devising policy recommendations to prevent global warming. Earth systems modeling requires knowledge of physics (atmosphere, oceans), chemistry (carbon cycle, soils), biology (decomposition, migration), engineering (renewable energy, carbon capture), economics (industry, energy use), human nature (stupidity, greed), and politics (even more stupidity, even more greed). As noted, the machine will have access to vast quantities of evidence to feed all these models. It will be able to suggest or carry out new experiments and expeditions to narrow down the inevitable uncertainties—for example, to discover the true extent of gas hydrates in shallow ocean reservoirs. It will be able to consider a vast range of possible policy recommendations—laws, nudges, markets, inventions, and geoengineering interventions—but of course it will also need to find ways to persuade us to go along with them.

The Limits of Superintelligence

While stretching your imagination, don't stretch it too far. A common mistake is to attribute godlike powers of omniscience to superintelligent AI systems—complete and perfect knowledge not just of the present but also of the future.[52] This is quite implausible because it requires an unphysical ability to determine the exact current state of the world as well as an unrealizable ability to simulate, much faster than real time, the operation of a world that includes the machine itself (not to mention billions of brains, which would still be the second-most-complex objects in the universe).

This is not to say that it is impossible to predict *some aspects* of the future with a reasonable degree of certainty—for example, I know what class I'll be teaching in what room at Berkeley almost a year from now, despite the protestations of chaos theorists about butterfly wings and all that. (Nor do I think that humans are anywhere close to predicting the future as well as the laws of physics allow!) Prediction depends on having the right abstractions—for example, I can predict that "I" will be "on stage in Wheeler Auditorium" on the Berkeley campus on the last Tuesday in April, but I cannot predict my exact location down to the millimeter or which atoms of carbon will have been incorporated into my body by then.

Machines are also subject to certain speed limits imposed by the real world on the rate at which new knowledge of the world can be acquired—one of the valid points made by Kevin Kelly in his article on oversimplified predictions about superhuman AI.[53] For example, to determine whether a specific drug cures a certain kind of cancer in an experimental animal, a scientist—human or machine—has two choices: inject the animal with the drug and wait several weeks or run a sufficiently accurate simulation. To run a simulation, however, requires a great deal of empirical knowledge of biology, some of which is currently unavailable; so, more model-building experiments would have to be done first. Undoubtedly, these would take time and must be done in the real world.

On the other hand, a machine scientist could run vast numbers of model-building experiments in parallel, could integrate their outcomes into an internally consistent (albeit very complex) model, and could compare the model's predictions with the entirety of experimental evidence known to biology. Moreover, simulating the model does not necessarily require a quantum-mechanical simulation of the entire organism down to the level of individual molecular reactions—which, as Kelly points out, would take more time than simply doing the experiment in the real world. Just as I can predict my future location on Tuesdays in April with some certainty, properties of biological systems can be predicted accurately with abstract models. (Among other reasons, this is because biology operates with robust control systems based on aggregate feedback loops, so that small variations in initial conditions usually don't lead to large variations in outcomes.) Thus, while *instantaneous* machine discoveries in the empirical sciences are unlikely, we can expect that science will proceed much faster with the help of machines. Indeed, it already is.

A final limitation of machines is that they are not human. This puts them at an intrinsic disadvantage when trying to model and predict one particular class of objects: humans. Our brains are all quite similar, so we can use them to simulate—to experience, if you will—the mental and emotional lives of others. This, for us, comes for free. (If you think about it, machines have an even greater advantage with each other: they can actually run each other's code!) For example, I don't need to be an expert on neural sensory systems to know what it feels like when *you* hit *your* thumb with a hammer. I can just hit *my* thumb with a hammer. Machines, on the other hand, have to start almost[54] from scratch in their understanding of humans: they have access only to our external behavior, plus all the neuroscience and psychology literature, and have to develop an understanding of how we work on that basis. In principle, they will be able to do this, but it's reasonable to suppose that acquiring a human-level or superhuman understanding of humans will take them longer than most other capabilities.

How Will AI Benefit Humans?

Our intelligence is responsible for our civilization. With access to greater intelligence we could have a greater—and perhaps far *better*—civilization. One can speculate about solving major open problems such as extending human life indefinitely or developing faster-than-light travel, but these staples of science fiction are not yet the driving force for progress in AI. (With superintelligent AI, we'll probably be able to invent all sorts of quasi-magical technologies, but it's hard to say now what those might be.) Consider, instead, a far more prosaic goal: raising the living standard of everyone on Earth, in a sustainable way, to a level that would be viewed as quite respectable in a developed country. Choosing (somewhat arbitrarily) *respectable* to mean the eighty-eighth percentile in the United States, the stated goal represents almost a tenfold increase in global gross domestic product (GDP), from $76 trillion to $750 trillion per year.[55]

To calculate the cash value of such a prize, economists use the *net present value* of the income stream, which takes into account the discounting of future income relative to the present. The extra income of $674 trillion per year has a net present value of roughly $13,500 trillion,[56] assuming a discount factor of 5 percent. So, in very crude terms, this is a ballpark figure for what human-level AI might be worth if it can deliver a respectable living standard for everyone. With numbers like this, it's not surprising that companies and countries are investing tens of billions of dollars annually in AI research and development.[57] Even so, the sums invested are minuscule compared to the size of the prize.

Of course, these are all made-up numbers unless one has some idea of *how* human-level AI could achieve the feat of raising living standards. It can do this only by increasing the per-capita production of goods and services. Put another way: the average human can never expect to consume more than the average human produces. The example of self-driving taxis discussed earlier in the chapter illustrates the multiplier effect of AI: with an automated service, it should be possible for (say) ten people to manage a fleet of one thousand vehicles, so each person is producing one hundred times as much transportation as before. The same goes for manufacturing the cars and for extracting the raw materials from which the cars are made. Indeed, some iron-ore mining operations in northern Australia, where temperatures regularly exceed 45 degrees Celsius (113 degrees Fahrenheit), are almost completely automated already.[58]

These present-day applications of AI are special-purpose systems: self-driving cars and self-operating mines have required huge investments in research, mechanical design, software engineering, and testing to develop the necessary algorithms and to make sure that they work as intended. That's just how things are done in all spheres of engineering. That's how things used to be done in personal travel too: if you wanted to travel from Europe to Australia and back in the seventeenth century, it would have involved a huge project costing vast sums of money, requiring years of planning, and carrying a high risk of death. Now we are used to the idea of transportation as a service (TaaS): if you need to be in Melbourne early next week, it just requires a few taps on your phone and a relatively minuscule amount of money.

General-purpose AI would be *everything as a service* (EaaS). There would be no need to employ armies of specialists in different disciplines, organized into hierarchies of contractors and subcontractors, in order to carry out a project. All embodiments of general-purpose AI would have access to all the knowledge and skills of the human race, and more besides. The only differentiation would be in the physical capabilities: dexterous legged robots for construction or surgery, wheeled robots for large-scale goods transportation, quadcopter robots for aerial inspections, and so on. In principle—politics and economics aside—everyone could have at their disposal an entire organization composed of software agents and physical robots, capable of designing and building bridges, improving crop yields, cooking dinner for a hundred guests, running elections, or doing whatever else needs doing. It's the *generality* of general-purpose intelligence that makes this possible.

History has shown, of course, that a tenfold increase in global GDP per capita is possible without AI—it's just that it took 190 years (from 1820 to 2010) to achieve that increase.[59] It required the development of factories, machine tools, automation, railways, steel, cars, airplanes, electricity, oil and gas production, telephones, radio, television, computers, the Internet, satellites, and many other revolutionary inventions. The tenfold increase in GDP posited in the preceding paragraphs is predicated not on further revolutionary technologies but on the ability of AI systems to employ what we already have more effectively and at greater scale.

Of course, there *will* be effects besides the purely material benefit of raising living standards. For example, personal tutoring is known to be far more effective than classroom teaching, but when done by humans it is simply unaffordable—and always will be—for the vast majority of people. With AI tutors, the potential of each child, no matter how poor, can be realized. The cost per child would be negligible, and that child would live a far richer and more productive life. The pursuit of artistic and intellectual endeavors, whether individually or collectively, would be a normal part of life rather than a rarefied luxury.

In the area of health, AI systems should enable researchers to unravel and master the vast complexities of human biology and thereby gradually banish disease. Greater insights into human psychology and neurochemistry should lead to broad improvements in mental health.

Perhaps more unconventionally, AI could enable far more effective authoring tools for virtual reality (VR) and could populate VR environments with far more interesting entities. This might turn VR into the medium of choice for literary and artistic expression, creating experiences of a richness and depth that is currently unimaginable.

And in the mundane world of daily life, an intelligent assistant and guide would—if well designed and not co-opted by economic and political interests—empower every individual to act effectively on their own behalf in an increasingly complex and sometimes hostile economic and political system. You would, in effect, have a high-powered lawyer, accountant, and political adviser on call at any time. Just as traffic jams are expected to be smoothed out by intermixing even a small percentage of autonomous vehicles, one can only hope that wiser policies and fewer conflicts will emerge from a better-informed and better-advised global citizenry.

These developments taken together could change the dynamic of history—at least that part of history that has been driven by conflicts within and between societies for access to the wherewithal of life. If the pie is essentially infinite, then fighting others for a larger share makes little sense. It would be like fighting over who gets the most digital copies of the newspaper—completely pointless when anyone can make as many digital copies as they want for free.

There are some limits to what AI can provide. The pies of land and raw materials are not infinite, so there cannot be unlimited population growth and not everyone will have a mansion in a private park. (This will eventually necessitate mining elsewhere in the solar system and constructing artificial habitats in space; but I promised not to talk about science fiction.) The pie of pride is also finite: only 1 percent of people can be in the top 1 percent on any given metric. If human happiness requires being in the top 1 percent, then 99 percent of humans are going to be unhappy, even when the bottom 1 percent has an objectively splendid lifestyle.[60] It will be important, then, for our cultures to gradually down-weight pride and envy as central elements of perceived self-worth.

As Nick Bostrom puts it at the end of his book *Superintelligence*, success in AI will yield "a civilizational trajectory that leads to a compassionate and jubilant use of humanity's cosmic endowment." If we fail to take advantage of what AI has to offer, we will have only ourselves to blame.

4

MISUSES OF AI

A compassionate and jubilant use of humanity's cosmic endowment sounds wonderful, but we also have to reckon with the rapid rate of innovation in the malfeasance sector. Ill-intentioned people are thinking up new ways to misuse AI so quickly that this chapter is likely to be outdated even before it attains printed form. Think of it not as depressing reading, however, but as a call to act before it is too late.

Surveillance, Persuasion, and Control

The automated Stasi

The Ministerium für Staatsicherheit of East Germany, more commonly known as the Stasi, is widely regarded as "one of the most effective and repressive intelligence and secret police agencies to have ever existed."[1] It maintained files on the great majority of East German households. It monitored phone calls, read letters, and planted hidden cameras in apartments and hotels. It was ruthlessly effective at identifying and eliminating dissident activity. Its preferred modus operandi was psychological destruction rather than imprisonment or execution. This level of control came at great cost, however: by some estimates, more than a quarter of working-age adults were Stasi informants. Stasi paper records have been estimated at twenty billion pages[2] and the task of processing and acting on the huge incoming flows of information began to exceed the capacity of any human organization.

It should come as no surprise, then, that intelligence agencies have spotted the potential for using AI in their work. For many years, they have been applying simple forms of AI technology, including voice recognition and identification of key words and phrases in both speech and text. Increasingly, AI systems are able to *understand the content* of what people are saying and doing, whether in speech, text, or video surveillance. In regimes where this technology is adopted for the purposes of control, it will be as if every citizen had their own personal Stasi operative watching over them twenty-four hours a day.[3]

Even in the civilian sphere, in relatively free countries, we are subject to increasingly effective surveillance. Corporations collect and sell information about our purchases, Internet and social network usage, electrical appliance usage, calling and texting records, employment, and health. Our locations can be tracked through our cell phones and our Internet-connected cars. Cameras recognize our faces on the street. All this data, and much more, can be pieced together by intelligent information integration systems to produce a fairly complete picture of what each of us is doing, how we live our lives, who we like and dislike, and how we will vote.[4] The Stasi will look like amateurs by comparison.

Controlling your behavior

Once surveillance capabilities are in place, the next step is to modify your behavior to suit those who are deploying this technology. One rather crude method is automated, personalized blackmail: a system that understands what you are doing—whether by listening, reading, or watching you—can easily spot things you should not be doing. Once it finds something, it will enter into correspondence with you to extract the largest possible amount of money (or to coerce behavior, if the goal is political control or espionage). The extraction of money works as the perfect reward signal for a reinforcement learning algorithm, so we can expect AI systems to improve rapidly in their ability to identify and profit from misbehavior. Early in 2015, I suggested to a computer security expert that automated blackmail systems, driven by reinforcement learning, might soon become feasible; he laughed and said it was already happening. The first blackmail bot to be widely publicized was Delilah, identified in July 2016.[5]

A more subtle way to change people's behavior is to modify their information environment so that they believe different things and make different decisions. Of course, advertisers have been doing this for centuries as a way of modifying the purchasing behavior of individuals. Propaganda as a tool of war and political domination has an even longer history.

So what's different now? First, because AI systems can track an individual's online reading habits, preferences, and likely state of knowledge, they can tailor specific messages to maximize impact on that individual while minimizing the risk that the information will be disbelieved. Second, the AI system knows whether the individual reads the message, how long they spend reading it, and whether they follow additional links within the message. It then uses these signals as immediate feedback on the success or failure of its attempt to influence each individual; in this way, it quickly learns to become more effective in its work. This is how content selection algorithms on social media have had their insidious effect on political opinions.

Another recent change is that the combination of AI, computer graphics, and speech synthesis is making it possible to generate *deepfakes*—realistic video and audio content of just about anyone saying or doing just about anything. The technology will require little more than a verbal description of the desired event, making it usable by more or less anyone in the world. Cell phone video of Senator X accepting a bribe from cocaine dealer Y at shady establishment Z? No problem! This kind of content can induce unshakeable beliefs in things that never happened.[6] In addition, AI systems can generate millions of false identities—the so-called bot armies—that can pump out billions of comments, tweets, and recommendations daily, swamping the efforts of mere humans to exchange truthful information. Online marketplaces such as eBay, Taobao, and Amazon that rely on reputation systems[7] to build trust between buyers and sellers are constantly at war with bot armies designed to corrupt the markets.

Finally, methods of control can be direct if a government is able to implement rewards and punishments based on behavior. Such a system treats people as reinforcement learning algorithms, training them to optimize the objective set by the state. The temptation for a government, particularly one with a top-down, engineering mind-set, is to reason as follows: it would be better if everyone behaved well, had a patriotic attitude, and contributed to the progress of the country; technology enables measurement of individual behavior, attitudes, and contributions; therefore, everyone will be better off if we set up a technology-based system of monitoring and control based on rewards and punishments.

There are several problems with this line of thinking. First, it ignores the psychic cost of living under a system of intrusive monitoring and coercion; outward harmony masking inner misery is hardly an ideal state. Every act of kindness ceases to be an act of kindness and becomes instead an act of personal score maximization and is perceived as such by the recipient. Or worse, the very concept of a voluntary act of kindness gradually becomes just a fading memory of something people used to do. Visiting an ailing friend in hospital will, under such a system, have no more moral significance and emotional value than stopping at a red light. Second, the scheme falls victim to the same failure mode as the standard model of AI, in that it assumes that the stated objective is in fact the true, underlying objective.

Inevitably, Goodhart's law will take over, whereby individuals optimize the official measure of outward behavior, just as universities have learned to optimize the "objective" measures of "quality" used by university ranking systems instead of improving their real (but unmeasured) quality.[8] Finally, the imposition of a uniform measure of behavioral virtue misses the point that a successful society may comprise a wide variety of individuals, each contributing in their own way.

A right to mental security

One of the great achievements of civilization has been the gradual improvement in physical security for humans. Most of us can expect to conduct our daily lives without constant fear of injury and death. Article 3 of the 1948 Universal Declaration of Human Rights states, "Everyone has the right to life, liberty and security of person."

I would like to suggest that everyone should also have the right to mental security—the right to live in a largely true information environment. Humans tend to believe the evidence of our eyes and ears. We trust our family, friends, teachers, and (some) media sources to tell us what they believe to be the truth. Even though we do not expect used-car salespersons and politicians to tell us the truth, we have trouble believing that they are lying as brazenly as they sometimes do. We are, therefore, extremely vulnerable to the technology of misinformation.

The right to mental security does not appear to be enshrined in the Universal Declaration. Articles 18 and 19 establish the rights of "freedom of thought" and "freedom of opinion and expression." One's thoughts and opinions are, of course, partly formed by one's information environment, which, in turn, is subject to Article 19's "right to . . . impart information and ideas through any media and regardless of frontiers." That is, anyone, anywhere in the world, has the right to impart false information to you. And therein lies the difficulty: democratic nations, particularly the United States, have for the most part been reluctant—or constitutionally unable—to prevent the imparting of false information on matters of public concern because of justifiable fears regarding government control of speech. Rather than pursuing the idea that there is no freedom of thought without access to true information, democracies seem to have placed a naïve trust in the idea that the truth will win out in the end, and this trust has left us unprotected. Germany is an exception; it recently passed the Network Enforcement Act, which requires content platforms to remove proscribed hate speech and fake news, but this has come under considerable criticism as being unworkable and undemocratic.[9]

For the time being, then, we can expect our mental security to remain under attack, protected mainly by commercial and volunteer efforts. These efforts include fact-checking sites such as factcheck.org and snopes.com—but of course other "fact-checking" sites are springing up to declare truth as lies and lies as truth.

The major information utilities such as Google and Facebook have come under extreme pressure in Europe and the United States to "do something about it." They are experimenting with ways to flag or relegate false content—using both AI and human screeners—and to direct users to verified sources that counteract the effects of misinformation. Ultimately, all such efforts rely on circular reputation systems, in the sense that sources are trusted because trusted sources report them to be trustworthy. If enough false information is propagated, these reputation systems can fail: sources that are actually trustworthy can become untrusted and vice versa, as appears to be occurring today with major media sources such as CNN and Fox News in the United States. Aviv Ovadya, a technologist working against misinformation, has called this the "infopocalypse—a catastrophic failure of the marketplace of ideas."[10]

One way to protect the functioning of reputation systems is to inject sources that are as close as possible to ground truth. A single fact that is *certainly true* can invalidate any number of sources that are only somewhat trustworthy, if those sources disseminate information contrary to the known fact. In many countries, notaries function as sources of ground truth to maintain the integrity of legal and real-estate information; they are usually disinterested third parties in any transaction and are licensed by governments or professional societies. (In the City of London, the Worshipful Company of Scriveners has been doing this since 1373, suggesting that a certain stability inheres in the role of truth telling.) If formal standards, professional qualifications, and licensing procedures emerge for fact-checkers, that would tend to preserve the validity of the information flows on which we depend. Organizations such as the W3C Credible Web group and the Credibility Coalition aim to develop technological and crowdsourcing methods for evaluating information providers, which would then allow users to filter out unreliable sources.

A second way to protect reputation systems is to impose a cost for purveying false information. Thus, some hotel rating sites accept reviews concerning a particular hotel only from those who have booked and paid for a room at that hotel through the site, while other rating sites accept reviews from anyone. It will come as no surprise that ratings at the former sites are far less biased, because they impose a cost (paying for an unnecessary hotel room) for fraudulent reviews.[11] *Regulatory* penalties are more controversial: no one wants a Ministry of Truth, and Germany's Network Enforcement Act penalizes only the content platform, not the person posting the fake news. On the other hand, just as many nations and many US states make it illegal to record telephone calls without permission, it ought, at least, to be possible to impose penalties for creating fictitious audio and video recordings of real people.

Finally, there are two other facts that work in our favor. First, almost no one actively wants, knowingly, to be lied to. (This is not to say that parents always inquire vigorously into the truthfulness of those who praise their children's intelligence and charm; it's just that they are less likely to seek such approval from someone who is known to lie at every opportunity.) This means that people of all political persuasions have an incentive to adopt tools that help them distinguish truth from lies. Second, no one wants to be known as a liar, least of all news outlets. This means that information providers—at least those for who reputation matters—have an incentive to join industry associations and subscribe to codes of conduct that favor truth telling. In turn, social media platforms can offer users the option of seeing content from only reputable sources that subscribe to these codes and subject themselves to third-party fact-checking.

Lethal Autonomous Weapons

The United Nations defines lethal autonomous weapons systems (AWS for short, because LAWS is quite confusing) as weapons systems that "locate, select, and eliminate human targets without human intervention." AWS have been described, with good reason, as the "third revolution in warfare," after gunpowder and nuclear weapons.

You may have read articles in the media about AWS; usually the article will call them *killer robots* and will be festooned with images from the Terminator movies. This is misleading in at least two ways: first, it suggests that autonomous weapons are a threat because they might take over the world and destroy the human race; second, it suggests that autonomous weapons will be humanoid, conscious, and evil.

The net effect of the media's portrayal of the issue has been to make it seem like science fiction. Even the German government has been taken in: it recently issued a statement[12] asserting that "having the ability to learn and develop self-awareness constitutes an indispensable attribute to be used to define individual functions or weapon systems as autonomous." (This makes as much sense as asserting that a missile isn't a missile unless it goes faster than the speed of light.) In fact, autonomous weapons will have the same degree of autonomy as a chess

 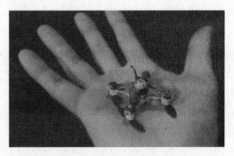

FIGURE 7: (left) Harop loitering weapon produced by Israel Aerospace Industries; (right) still image from the *Slaughterbots* video showing a possible design for an autonomous weapon containing a small, explosive-driven projectile.

program, which is given the mission of winning the game but decides by itself where to move its pieces and which enemy pieces to eliminate.

AWS are not science fiction. They already exist. Probably the clearest example is Israel's Harop (figure 7, left), a *loitering munition* with a ten-foot wingspan and a fifty-pound warhead. It searches for up to six hours in a given geographical region for any target that meets a given criterion and then destroys it. The criterion could be "emits a radar signal resembling antiaircraft radar" or "looks like a tank."

By combining recent advances in miniature quadrotor design, miniature cameras, computer vision chips, navigation and mapping algorithms, and methods for detecting and tracking humans, it would be possible in fairly short order to field an antipersonnel weapon like the Slaughterbot[13] shown in figure 7 (right). Such a weapon could be tasked with attacking anyone meeting certain visual criteria (age, gender, uniform, skin color, and so on) or even specific individuals based on face recognition. I'm told that the Swiss Defense Department has already built and tested a real Slaughterbot and found that, as expected, the technology is both feasible and lethal.

Since 2014, diplomatic discussions have been underway in Geneva that may lead to a treaty banning AWS. At the same time, some of the major participants in these discussions (the United States, China,

Russia, and to some extent Israel and the UK) are engaged in a dangerous competition to develop autonomous weapons. In the United States,
for example, the CODE (Collaborative Operations in Denied Environments) program aims to move towards autonomy by enabling drones to
function with at best intermittent radio contact. The drones will "hunt
in packs, like wolves" according to the program manager.[14] In 2016, the
US Air Force demonstrated the in-flight deployment of 103 Perdix
micro-drones from three F/A-18 fighters. According to the announcement, "Perdix are not pre-programmed synchronized individuals, they
are a collective organism, sharing one distributed brain for decision-
making and adapting to each other like swarms in nature."[15]

You may think it's pretty obvious that building machines that can
decide to kill humans is a bad idea. But "pretty obvious" is not always
persuasive to governments—including some of those listed in the preceding paragraph—who are bent on achieving what they think of as
strategic superiority. A more convincing reason to reject autonomous
weapons is that they are *scalable weapons of mass destruction*.

Scalable is a term from computer science; a process is scalable if
you can do a million times more of it essentially by buying a million
times more hardware. Thus, Google handles roughly five billion
search requests per day by having not millions of employees but millions of computers. With autonomous weapons, you can do a million
times more killing by buying a million times more weapons, *precisely because the weapons are autonomous*. Unlike remotely piloted
drones or AK-47s, they don't need individual human supervision to do
their work.

As weapons of mass destruction, scalable autonomous weapons
have advantages for the attacker compared to nuclear weapons and
carpet bombing: they leave property intact and can be applied selec-
tively to eliminate only those who might threaten an occupying force.
They could certainly be used to wipe out an entire ethnic group or all
the adherents of a particular religion (if adherents have visible indicia).
Moreover, whereas the use of nuclear weapons represents a cataclys-

mic threshold that we have (often by sheer luck) avoided crossing since 1945, there is no such threshold with scalable autonomous weapons. Attacks could escalate smoothly from one hundred casualties to one thousand to ten thousand to one hundred thousand. In addition to actual attacks, the mere *threat* of attacks by such weapons makes them an effective tool for terror and oppression. Autonomous weapons will greatly reduce human security at all levels: personal, local, national, and international.

This is not to say that autonomous weapons will be the end of the world in the way envisaged in the Terminator movies. They need not be especially intelligent—a self-driving car probably needs to be smarter—and their missions will not be of the "take over the world" variety. The existential risk from AI does not come primarily from simple-minded killer robots. On the other hand, superintelligent machines in conflict with humanity could certainly arm themselves this way, by turning relatively stupid killer robots into physical extensions of a global control system.

Eliminating Work as We Know It

Thousands of media articles and opinion pieces and several books have been written on the topic of robots taking jobs from humans. Research centers are springing up all over the world to understand what is likely to happen.[16] The titles of Martin Ford's *Rise of the Robots: Technology and the Threat of a Jobless Future*[17] and Calum Chace's *The Economic Singularity: Artificial Intelligence and the Death of Capitalism*[18] do a pretty good job of summarizing the concern. Although, as will soon become evident, I am by no means qualified to opine on what is essentially a matter for economists,[19] I suspect that the issue is too important to leave entirely to them.

The issue of *technological unemployment* was brought to the fore in a famous article, "Economic Possibilities for Our Grandchildren," by

John Maynard Keynes. He wrote the article in 1930, when the Great Depression had created mass unemployment in Britain, but the topic has a much longer history. Aristotle, in Book I of his *Politics*, presents the main point quite clearly:

> For if every instrument could accomplish its own work, obeying or anticipating the will of others . . . if, in like manner, the shuttle would weave and the plectrum touch the lyre without a hand to guide them, chief workmen would not want servants, nor masters slaves.

Everyone agrees with Aristotle's observation that there is an immediate reduction in employment when an employer finds a mechanical method to perform work previously done by a person. The issue is whether the so-called compensation effects that ensue—and that tend to increase employment—will eventually make up for this reduction. The optimists say yes—and in the current debate, they point to all the new jobs that emerged after previous industrial revolutions. The pessimists say no—and in the current debate, they argue that machines will do all the "new jobs" too. When a machine replaces one's physical labor, one can sell mental labor. When a machine replaces one's mental labor, what does one have left to sell?

In *Life 3.0*, Max Tegmark depicts the debate as a conversation between two horses discussing the rise of the internal combustion engine in 1900. One predicts "new jobs for horses. . . . That's what's always happened before, like with the invention of the wheel and the plow." For most horses, alas, the "new job" was to be pet food.

The debate has persisted for millennia because there are effects in both directions. The actual outcome depends on which effects matter more. Consider, for example, what happens to housepainters as technology improves. For the sake of simplicity, I'll let the width of the paintbrush stand for the degree of automation:

- If the brush is one hair (a tenth of a millimeter) wide, it takes thousands of person-years to paint a house and essentially no housepainters are employed.

- With brushes a millimeter wide, perhaps a few delicate murals are painted in the royal palace by a handful of painters. At one centimeter, the nobility begin to follow suit.

- At ten centimeters (four inches), we reach the realm of practicality: most homeowners have their houses painted inside and out, although perhaps not all that frequently, and thousands of housepainters find jobs.

- Once we get to wide rollers and spray guns—the equivalent of a paintbrush about a meter wide—the price goes down considerably, but demand may begin to saturate so the number of housepainters drops somewhat.

- When one person manages a team of one hundred housepainting robots—the productivity equivalent of a paintbrush one hundred meters wide—then whole houses can be painted in an hour and very few housepainters will be working.

Thus, the *direct* effects of technology work both ways: at first, by increasing productivity, technology can increase employment by reducing the price of an activity and thereby increasing demand; subsequently, further increases in technology mean that fewer and fewer humans are required. Figure 8 illustrates these developments.[20]

Many technologies exhibit similar curves. If, in some given sector of the economy, we are to the left of the peak, then improving technology increases employment in that sector; present-day examples might include tasks such as graffiti removal, environmental cleanup, inspection of shipping containers, and housing construction in less developed countries, all of which might become more economically feasible if we have robots to help us. If we are already to the right of the peak, then further automation decreases employment. For example,

number of
housepainters employed

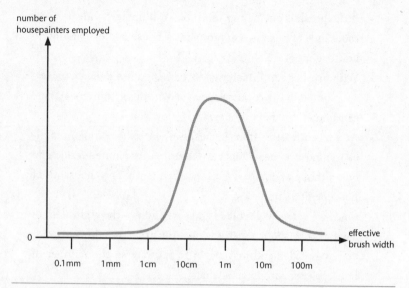

effective
brush width

0.1mm   1mm   1cm   10cm   1m   10m   100m

FIGURE 8: A notional graph of housepainting employment as painting technology improves.

it's not hard to predict that elevator operators will continue to be squeezed out. In the long run, we have to expect that most industries are going to be pushed to the far right on the curve. One recent article, based on a careful econometric study by economists David Autor and Anna Salomons, states that "over the last 40 years, jobs have fallen in every single industry that introduced technologies to enhance productivity."[21]

What about the compensation effects described by the economic optimists?

- Some people have to make the painting robots. How many? Far *fewer* than the number of housepainters the robots replace—otherwise, it would cost more to paint houses with robots, not less, and no one would buy the robots.

- Housepainting becomes somewhat cheaper, so people call in the housepainters a bit more often.
- Finally, because we pay less for housepainting, we have more money to spend on other things, thereby increasing employment in other sectors.

Economists have tried to measure the size of these effects in various industries experiencing increased automation, but the results are generally inconclusive.

Historically, most mainstream economists have argued from the "big picture" view: automation increases productivity, so, *as a whole*, humans are better off, in the sense that we enjoy more goods and services for the same amount of work.

Economic theory does not, unfortunately, predict that *each* human will be better off as a result of automation. Generally, automation increases the share of income going to capital (the owners of the housepainting robots) and decreases the share going to labor (the ex-housepainters). The economists Erik Brynjolfsson and Andrew McAfee, in *The Second Machine Age*, argue that this has already been happening for several decades. Data for the United States are shown in figure 9. They indicate that between 1947 and 1973, wages and productivity increased together, but after 1973, wages stagnated even while productivity roughly doubled. Brynjolfsson and McAfee call this the *Great Decoupling*. Other leading economists have also sounded the alarm, including Nobel laureates Robert Shiller, Mike Spence, and Paul Krugman; Klaus Schwab, head of the World Economic Forum; and Larry Summers, former chief economist of the World Bank and Treasury secretary under President Bill Clinton.

Those arguing against the notion of technological unemployment often point to bank tellers, whose work can be done in part by ATMs, and retail cashiers, whose work is sped up by barcodes and RFID tags on merchandise. It is often claimed that these occupations are growing

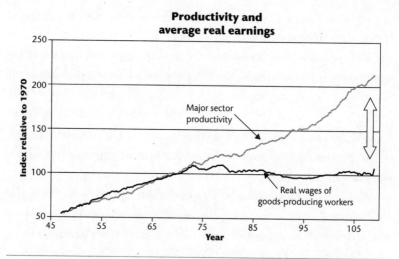

FIGURE 9: Economic production and real median wages in the United States since 1947. (Data from the Bureau of Labor Statistics.)

because of technology. Indeed, the number of tellers in the United States roughly doubled from 1970 to 2010, although it should be noted that the US population grew by 50 percent and the financial sector by over 400 percent in the same period,[22] so it is difficult to attribute all, or perhaps any, of the employment growth to ATMs. Unfortunately, between 2010 and 2016 about one hundred thousand tellers lost their jobs, and the US Bureau of Labor Statistics (BLS) predicts another forty thousand job losses by 2026: "Online banking and automation technology are expected to continue replacing more job duties that tellers traditionally performed."[23] The data on retail cashiers are no more encouraging: the number per capita dropped by 5 percent from 1997 to 2015, and the BLS says, "Advances in technology, such as self-service checkout stands in retail stores and increasing online sales, will continue to limit the need for cashiers." Both sectors appear to be on the downslope. The same is true of almost all low-skilled occupations that involve working with machines.

Which occupations are about to decline as new, AI-based technology arrives? The prime example cited in the media is that of driving. In the United States there are about 3.5 million truck drivers; many of these jobs would be vulnerable to automation. Amazon, among other companies, is already using self-driving trucks for freight haulage on interstate freeways, albeit currently with human backup drivers.[24] It seems very likely that the long-haul part of each truck journey will soon be autonomous, while humans, for the time being, will handle city traffic, pickup, and delivery. As a consequence of these expected developments, very few young people are interested in trucking as a career; ironically, there is currently a significant shortage of truck drivers in the United States, which is only hastening the onset of automation.

White-collar jobs are also at risk. For example, the BLS projects a 13 percent decline in per-capita employment of insurance underwriters from 2016 to 2026: "Automated underwriting software allows workers to process applications more quickly than before, reducing the need for as many underwriters." If language technology develops as expected, many sales and customer service jobs will also be vulnerable, as well as jobs in the legal profession. (In a 2018 competition, AI software outscored experienced law professors in analyzing standard nondisclosure agreements and completed the task two hundred times faster.[25]) Routine forms of computer programming—the kind that is often outsourced today—are also likely to be automated. Indeed, almost anything that can be outsourced is a good candidate for automation, because outsourcing involves decomposing jobs into tasks that can be parceled up and distributed in a decontextualized form. The *robot process automation* industry produces software tools that achieve exactly this effect for clerical tasks performed online.

As AI progresses, it is certainly possible—perhaps even likely—that within the next few decades essentially all routine physical and mental labor will be done more cheaply by machines. Since we ceased to be hunter-gatherers thousands of years ago, our societies have used most people as robots, performing repetitive manual and mental tasks, so it is perhaps not surprising that robots will soon take on these roles.

When this happens, it will push wages below the poverty line for those people who are unable to compete for the highly skilled jobs that remain. Larry Summers put it this way: "It may well be that, given the possibilities for substitution [of capital for labor], some categories of labor will not be able to earn a subsistence income."[26] This is precisely what happened to the horses: mechanical transportation became cheaper than the upkeep cost of a horse, so horses became pet food. Faced with the socioeconomic equivalent of becoming pet food, humans will be rather unhappy with their governments.

Faced with potentially unhappy humans, governments around the world are beginning to devote some attention to the issue. Most have already discovered that the idea of retraining everyone as a data scientist or robot engineer is a nonstarter—the world might need five or ten million of these, but nowhere close to the billion or so jobs that are at risk. Data science is a very tiny lifeboat for a giant cruise ship.[27]

Some are working on "transition plans"—but transition to what? We need a plausible destination in order to plan a transition—that is, we need a plausible picture of a desirable future economy where most of what we currently call work is done by machines.

One rapidly emerging picture is that of an economy where far fewer people work because work is unnecessary. Keynes envisaged just such a future in his essay "Economic Possibilities for Our Grandchildren." He described the high unemployment afflicting Great Britain in 1930 as a "temporary phase of maladjustment" caused by an "increase of technical efficiency" that took place "faster than we can deal with the problem of labour absorption." He did not, however, imagine that in the long run—after a century of further technological advances—there would be a return to full employment:

> Thus for the first time since his creation man will be faced with
> his real, his permanent problem—how to use his freedom from
> pressing economic cares, how to occupy the leisure, which science

and compound interest will have won for him, to live wisely and agreeably and well.

Such a future requires a radical change in our economic system, because, in many countries, those who do not work face poverty or destitution. Thus, modern proponents of Keynes's vision usually support some form of *universal basic income*, or UBI. Funded by value-added taxes or by taxes on income from capital, UBI would provide a reasonable income to every adult, regardless of circumstance. Those who aspire to a higher standard of living can still work without losing the UBI, while those who do not can spend their time as they see fit. Perhaps surprisingly, UBI has support across the political spectrum, ranging from the Adam Smith Institute[28] to the Green Party.[29]

For some, UBI represents a version of paradise.[30] For others, it represents an admission of failure—an assertion that most people will have nothing of economic value to contribute to society. They can be fed and housed—mostly by machines—but otherwise left to their own devices. The truth, as always, lies somewhere in between, and it depends largely on how one views human psychology. Keynes, in his essay, made a clear distinction between those who strive and those who enjoy—those "purposive" people for whom "jam is not jam unless it is a case of jam to-morrow and never jam to-day" and those "delightful" people who are "capable of taking direct enjoyment in things." The UBI proposal assumes that the great majority of people are of the delightful variety.

Keynes suggests that striving is one of the "habits and instincts of the ordinary man, bred into him for countless generations" rather than one of the "real values of life." He predicts that this instinct will gradually disappear. Against this view, one may suggest that striving is intrinsic to what it means to be truly human. Rather than striving and enjoying being mutually exclusive, they are often inseparable: true enjoyment and lasting fulfillment come from having a purpose and achieving it (or at least trying), usually in the face of obstacles, rather than from passive consumption of immediate pleasure. There is a difference between climbing Everest and being deposited on top by helicopter.

The connection between striving and enjoying is a central theme for our understanding of how to fashion a desirable future. Perhaps future generations will wonder why we ever worried about such a futile thing as "work." Just in case that change in attitudes is slow in coming, let's consider the economic implications of the view that most people will be better off with something useful to do, even though the great majority of goods and services will be produced by machines with very little human supervision. Inevitably, most people will be engaged in supplying interpersonal services that can be provided—or which we *prefer* to be provided—only by humans. That is, if we can no longer supply routine physical labor and routine mental labor, we can still supply our humanity. We will need to become good at being human.[31]

Current professions of this kind include psychotherapists, executive coaches, tutors, counselors, companions, and those who care for children and the elderly. The phrase *caring professions* is often used in this context, but that is misleading: it has a positive connotation for those providing care, to be sure, but a negative connotation of dependency and helplessness for the recipients of care. But consider this observation, again from Keynes:

> It will be those peoples, who can keep alive, and cultivate into a fuller perfection, the art of life itself and do not sell themselves for the means of life, who will be able to enjoy the abundance when it comes.

All of us need help in learning "the art of life itself." This is not a matter of dependency but of growth. The capacity to inspire others and to confer the ability to appreciate and to create—be it in art, music, literature, conversation, gardening, architecture, food, wine, or video games—is likely to be more needed than ever.

The next question is income distribution. In most countries, this has been moving in the wrong direction for several decades. It's a complex issue, but one thing is clear: high incomes and high social standing usually follow from providing high added value. The profession of childcare, to pick one example, is associated with low incomes and low social standing. This is, in part, a consequence of the fact that we don't really know how to do it. Some practitioners are naturally good at it, but many are not. Contrast this with, say, orthopedic surgery. We wouldn't just hire bored teenagers who need a bit of spare cash and put them to work as orthopedic surgeons at five dollars an hour plus all they can eat from the fridge. We have put centuries of research into understanding the human body and how to fix it when it's broken, and practitioners must undergo years of training to learn all this knowledge and the skills necessary to apply it. As a result, orthopedic surgeons are highly paid and highly respected. They are highly paid not just because they know a lot and have a lot of training but also because all that knowledge and training actually works. It enables them to add a great deal of value to other people's lives—especially people with broken bits.

Unfortunately, our scientific understanding of the mind is shockingly weak and our scientific understanding of happiness and fulfillment is even weaker. We simply don't know how to add value to each other's lives in consistent, predictable ways. We have had moderate success with certain psychiatric disorders, but we are still fighting a Hundred Years' Literacy War over something as basic as teaching children to read.[32] We need a radical rethinking of our educational system and our scientific enterprise to focus more attention on the human rather than the physical world. (Joseph Aoun, president of Northeastern University, argues that universities should be teaching and studying "humanics."[33]) It sounds odd to say that happiness should be an engineering discipline, but that seems to be the inevitable conclusion.

Such a discipline would build on basic science—a better understanding of how human minds work at the cognitive and emotional levels— and would train a wide variety of practitioners, ranging from life architects, who help individuals plan the overall shape of their life trajectories, to professional experts in topics such as curiosity enhancement and personal resilience. If based on real science, these professions need be no more woo-woo than bridge designers and orthopedic surgeons are today.

Reworking our education and research institutions to create this basic science and to convert it into training programs and credentialed professions will take decades, so it's a good idea to start now and a pity we didn't start long ago. The final result—if it works—would be a world well worth living in. Without such a rethinking, we risk an unsustainable level of socioeconomic dislocation.

Usurping Other Human Roles

We should think twice before allowing machines to take over roles involving interpersonal services. If being human is our main selling point to other humans, so to speak, then making imitation humans seems like a bad idea. Fortunately for us, we have a distinct advantage over machines when it comes to knowing how other humans feel and how they will react. Nearly every human knows what it's like to hit one's thumb with a hammer or to feel unrequited love.

Counteracting this natural human advantage is a natural human disadvantage: the tendency to be fooled by appearances—especially human appearances. Alan Turing warned against making robots resemble humans:[34]

> I certainly hope and believe that no great efforts will be put into making machines with the most distinctively human, but non-intellectual, characteristics such as the shape of the human body;

it appears to me quite futile to make such attempts and their results would have something like the unpleasant quality of artificial flowers.

Unfortunately, Turing's warning has gone unheeded. Several research groups have produced eerily lifelike robots, as shown in figure 10.

As research tools, the robots may provide insights into how humans interpret robot behavior and communication. As prototypes for future commercial products, they represent a form of dishonesty. They bypass our conscious awareness and appeal directly to our emotional selves, perhaps convincing us that they are endowed with real intelligence. Imagine, for example, how much easier it would be to switch off and recycle a squat, gray box that was malfunctioning—even if it was squawking about not wanting to be switched off—than it would be to do the same for JiaJia or Geminoid DK. Imagine also how confusing and perhaps psychologically disturbing it would be for babies and small children to be cared for by entities that appear to be human, like their parents, but are somehow not; that appear to care about them, like their parents, but in fact do not.

FIGURE 10: (left) JiaJia, a robot built at the University of Science and Technology of China; (right) Geminoid DK, a robot designed by Hiroshi Ishiguro at Osaka University in Japan and modeled on Henrik Schärfe of Aalborg University in Denmark.

Beyond a basic capability to convey nonverbal information via facial expression and movement—which even Bugs Bunny manages to do with ease—there is no good reason for robots to have humanoid form. There are also good, practical reasons *not* to have humanoid form—for example, our bipedal stance is relatively unstable compared to quadrupedal locomotion. Dogs, cats, and horses fit into our lives well, and their physical form is a very good clue as to how they are likely to behave. (Imagine if a horse suddenly started behaving like a dog!) The same should be true of robots. Perhaps a four-legged, two-armed, centaur-like morphology would be a good standard. An accurately humanoid robot makes as much sense as a Ferrari with a top speed of five miles per hour or a "raspberry" ice-cream cone made from beetroot-tinted cream of chopped liver.

The humanoid aspect of some robots has already contributed to political as well as emotional confusion. On October 25, 2017, Saudi Arabia granted citizenship to Sophia, a humanoid robot that has been described as little more than "a chatbot with a face"[35] and worse.[36] Perhaps this was a public relations stunt, but a proposal emanating from the European Parliament's Committee on Legal Affairs is entirely serious.[37] It recommends

> creating a specific legal status for robots in the long run, so that at least the most sophisticated autonomous robots could be established as having the status of electronic persons responsible for making good any damage they may cause.

In other words, the *robot itself* would be legally responsible for damage, rather than the owner or manufacturer. This implies that robots will own financial assets and be subject to sanctions if they do not comply. Taken literally, this does not make sense. For example, if we were to imprison the robot for nonpayment, why would it care?

In addition to the needless and even absurd elevation of the status of robots, there is a danger that the increased use of machines in

FIGURE 11: Max (Matt Damon) meets his parole officer in *Elysium*.

decisions affecting people will degrade the status and dignity of humans. This possibility is illustrated perfectly in a scene from the science-fiction movie *Elysium*, when Max (Matt Damon) pleads his case before his "parole officer" (figure 11) to explain why the extension of his sentence is unjustified. Needless to say, Max is unsuccessful. The parole officer even chides him for failing to display a suitably deferential attitude.

One can think of such an assault on human dignity in two ways. The first is obvious: by giving machines authority over humans, we relegate ourselves to a second-class status and lose the right to participate in decisions that affect us. (A more extreme form of this is giving machines the authority to kill humans, as discussed earlier in the chapter.) The second is indirect: even if you believe it is not the *machines* making the decision but *those humans who designed and commissioned the machines*, the fact that those human designers and commissioners do not consider it worthwhile to weigh the individual circumstances of each human subject in such cases suggests that they attach little value to the lives of others. This is perhaps a symptom of the beginning of a great separation between an elite served by humans and a vast underclass served, and controlled, by machines.

In the EU, Article 22 of the 2018 General Data Protection Regulation, or GDPR, explicitly forbids the granting of authority to machines in such cases:

> The data subject shall have the right not to be subject to a decision based solely on automated processing, including profiling, which produces legal effects concerning him or her or similarly significantly affects him or her.

Although this sounds admirable in principle, it remains to be seen—at least at the time of writing—how much impact this will have in practice. It is often so much easier, faster, and cheaper to leave the decisions to the machine.

One reason for all the concern about automated decisions is the potential for *algorithmic bias*—the tendency of machine learning algorithms to produce inappropriately biased decisions about loans, housing, jobs, insurance, parole, sentencing, college admission, and so on. The explicit use of criteria such as race in these decisions has been illegal for decades in many countries and is prohibited by Article 9 of the GDPR for a very wide range of applications. That does not mean, of course, that by excluding race from the data we necessarily get racially unbiased decisions. For example, beginning in the 1930s, the government-sanctioned practice of redlining caused certain zip codes in the United States to be off-limits for mortgage lending and other forms of investment, leading to declining real-estate values. It just so happened that those zip codes were largely populated by African Americans.

To prevent redlining, now only the first three digits of the five-digit zip code can be used in making credit decisions. In addition, the decision process must be amenable to inspection, to ensure no other "accidental" biases are creeping in. The EU's GDPR is often said to provide a general "right to an explanation" for any automated decision,[38] but the actual language of Article 14 merely requires

> meaningful information about the logic involved, as well as the significance and the envisaged consequences of such processing for the data subject.

At present, it is unknown how courts will enforce this clause. It's possible that the hapless consumer will just be handed a description of the particular deep learning algorithm used to train the classifier that made the decision.

Nowadays, the likely causes of algorithmic bias lie in the data rather than in the deliberate malfeasance of corporations. In 2015, *Glamour* magazine reported a disappointing finding: "The first female Google image search result for 'CEO' appears TWELVE rows down—and it's Barbie." (There were some actual women in the 2018 results, but most of them were models portraying CEOs in generic stock photos, rather than actual female CEOs; the 2019 results are somewhat better.) This is a consequence not of deliberate gender bias in Google's image search ranking but of preexisting bias in the culture that produces the data: there are far more male than female CEOs, and when people want to depict an "archetypal" CEO in a captioned image, they almost always pick a male figure. The fact that the bias lies primarily in the data does not, of course, mean that there is no obligation to take steps to counteract the problem.

There are other, more technical reasons why the naïve application of machine learning methods can produce biased outcomes. For example, minorities are, by definition, less well represented in population-wide data samples; hence, predictions for individual members of minorities may be less accurate if such predictions are made largely on the basis of data from other members of the same group. Fortunately, a good deal of attention has been paid to the problem of removing inadvertent bias from machine learning algorithms, and there are now methods that produce unbiased results according to several plausible and desirable definitions of fairness.[39] The mathematical analysis of these definitions of fairness shows that they cannot be achieved simultaneously and that, when enforced, they result in lower prediction accuracy and, in the case of lending decisions, lower profit for the lender. This is perhaps disappointing, but at least it makes clear the trade-offs involved in avoiding algorithmic bias. One hopes that awareness of these methods and of the issue itself will spread quickly among policy makers, practitioners, and users.

If handing authority over individual humans to machines is sometimes problematic, what about authority over lots of humans? That is, should we put machines in political and management roles? At present this may seem far-fetched. Machines cannot sustain an extended conversation and lack the basic understanding of the factors that are relevant to making decisions with broad scope, such as whether to raise the minimum wage or to reject a merger proposal from another corporation. The trend, however, is clear: machines are making decisions at higher and higher levels of authority in many areas. Take airlines, for example. First, computers helped in the construction of flight schedules. Soon, they took over allocation of flight crews, the booking of seats, and the management of routine maintenance. Next, they were connected to global information networks to provide real-time status reports to airline managers, so that managers could cope with disruption effectively. Now they are taking over the job of managing disruption: rerouting planes, rescheduling staff, rebooking passengers, and revising maintenance schedules.

This is all to the good from the point of view of airline economics and passenger experience. The question is whether the computer system remains a tool of humans, or humans become tools of the computer system—supplying information and fixing bugs when necessary, but no longer understanding in any depth how the whole thing is working. The answer becomes clear when the system goes down and global chaos ensues until it can be brought back online. For example, a single "computer glitch" on April 3, 2018, caused fifteen thousand flights in Europe to be significantly delayed or canceled.[40] When trading algorithms caused the 2010 "flash crash" on the New York Stock Exchange, wiping out $1 trillion in a few minutes, the only solution was to shut down the exchange. What happened is still not well understood.

Before there was any technology, human beings lived, like most animals, hand to mouth. We stood directly on the ground, so to speak. Technology gradually raised us up on a pyramid of machinery, increasing our footprint as individuals and as a species. There are different ways we can design the relationship between humans and machines. If we design it so that humans retain sufficient understanding, authority, and autonomy, the technological parts of the system can greatly magnify human capabilities, allowing each of us to stand on a vast pyramid of capabilities—a demigod, if you like. But consider the worker in an online-shopping fulfillment warehouse. She is more productive than her predecessors because she has a small army of robots bringing her storage bins to pick items from; but she is a part of a larger system controlled by intelligent algorithms that decide where she should stand and which items she should pick and dispatch. She is already partly buried in the pyramid, not standing on top of it. It's only a matter of time before the sand fills the spaces in the pyramid and her role is eliminated.

OVERLY INTELLIGENT AI

The Gorilla Problem

It doesn't require much imagination to see that making something smarter than yourself could be a bad idea. We understand that our control over our environment and over other species is a result of our intelligence, so the thought of something else being more intelligent than us—whether it's a robot or an alien—immediately induces a queasy feeling.

Around ten million years ago, the ancestors of the modern gorilla created (accidentally, to be sure) the genetic lineage leading to modern humans. How do the gorillas feel about this? Clearly, if they were able to tell us about their species' current situation vis-à-vis humans, the consensus opinion would be very negative indeed. Their species has essentially no future beyond that which we deign to allow. We do not want to be in a similar situation vis-à-vis superintelligent machines. I'll call this the *gorilla problem*—specifically, the problem of whether humans can maintain their supremacy and autonomy in a world that includes machines with substantially greater intelligence.

Charles Babbage and Ada Lovelace, who designed and wrote programs for the Analytical Engine in 1842, were aware of its potential but seemed to have no qualms about it.[1] In 1847, however, Richard Thornton, editor of the *Primitive Expounder*, a religious journal, railed against mechanical calculators:[2]

> Mind . . . outruns itself and does away with the necessity of its own existence by inventing machines to do its own *thinking*. . . . But who knows that such machines when brought to greater perfection, may not think of a plan to remedy all their own defects and then grind out ideas beyond the ken of mortal mind!

This is perhaps the first speculation concerning existential risk from computing devices, but it remained in obscurity.

In contrast, Samuel Butler's novel *Erewhon*, published in 1872, developed the theme in far greater depth and achieved immediate success. Erewhon is a country in which all mechanical devices have been banned after a terrible civil war between the machinists and anti-machinists. One part of the book, called "The Book of the Machines," explains the origins of this war and presents the arguments of both sides.[3] It is eerily prescient of the debate that has re-emerged in the early years of the twenty-first century.

The anti-machinists' main argument is that machines will advance to the point where humanity loses control:

> Are we not ourselves creating our successors in the supremacy of the earth? Daily adding to the beauty and delicacy of their organization, daily giving them greater skill and supplying more and more of that self-regulating self-acting power which will be better than any intellect? . . . In the course of ages we shall find ourselves the inferior race. . . .
>
> We must choose between the alternative of undergoing much present suffering, or seeing ourselves gradually superseded by our own creatures, till we rank no higher in comparison with them,

than the beasts of the field with ourselves. . . . Our bondage will
steal upon us noiselessly and by imperceptible approaches.

The narrator also relates the pro-machinists' principal counter-
argument, which anticipates the man–machine symbiosis argument
that we will explore in the next chapter:

> There was only one serious attempt to answer it. Its author said
> that machines were to be regarded as a part of man's own physical
> nature, being really nothing but extra-corporeal limbs.

Although the anti-machinists in Erewhon win the argument, Butler
himself appears to be of two minds. On the one hand, he complains
that "Erewhonians are . . . quick to offer up common sense at the
shrine of logic, when a philosopher arises among them, who carries
them away through his reputation for especial learning" and says,
"They cut their throats in the matter of machinery." On the other
hand, the Erewhonian society he describes is remarkably harmonious,
productive, and even idyllic. The Erewhonians fully accept the folly
of re-embarking on the course of mechanical invention, and regard
those remnants of machinery kept in museums "with the feelings of
an English antiquarian concerning Druidical monuments or flint arrow heads."

Butler's story was evidently known to Alan Turing, who considered the long-term future of AI in a lecture given in Manchester
in 1951:[4]

> It seems probable that once the machine thinking method had
> started, it would not take long to outstrip our feeble powers. There
> would be no question of the machines dying, and they would
> be able to converse with each other to sharpen their wits. At some
> stage therefore we should have to expect the machines to take
> control, in the way that is mentioned in Samuel Butler's *Erewhon*.

In the same year, Turing repeated these concerns in a radio lecture broadcast throughout the UK on the BBC Third Programme:

> If a machine can think, it might think more intelligently than we do, and then where should we be? Even if we could keep the machines in a subservient position, for instance by turning off the power at strategic moments, we should, as a species, feel greatly humbled. . . . This new danger . . . is certainly something which can give us anxiety.

When the Erewhonian anti-machinists "feel seriously uneasy about the future," they see it as their "duty to check the evil while we can still do so," and they destroy all the machines. Turing's response to the "new danger" and "anxiety" is to consider "turning off the power" (although it will be clear shortly that this is not really an option). In Frank Herbert's classic science-fiction novel *Dune*, set in the far future, humanity has barely survived the Butlerian Jihad, a cataclysmic war with the "thinking machines." A new commandment has emerged: *"Thou shalt not make a machine in the likeness of a human mind."* This commandment precludes computing devices of any kind.

All these drastic responses reflect the inchoate fears that machine intelligence evokes. Yes, the prospect of superintelligent machines does make one uneasy. Yes, it is logically possible that such machines could take over the world and subjugate or eliminate the human race. If that is all one has to go on, then indeed the only plausible response available to us, at the present time, is to attempt to curtail artificial intelligence research—specifically, to ban the development and deployment of general-purpose, human-level AI systems.

Like most other AI researchers, I recoil at this prospect. How dare anyone tell me what I can and cannot think about? Anyone proposing an end to AI research is going to have to do a lot of convincing. Ending AI research would mean forgoing not just one of the principal avenues for understanding how human intelligence works but also a golden opportunity to improve the human condition—to make a far better civilization. The economic value of human-level AI is measurable in the thousands of trillions of dollars, so the momentum behind AI research from corporations and governments is likely to be enormous. It will overwhelm the vague objections of a philosopher, no matter how great his or her "reputation for especial learning," as Butler puts it.

A second drawback to the idea of banning general-purpose AI is that it's a difficult thing to ban. Progress on general-purpose AI occurs primarily on the whiteboards of research labs around the world, as mathematical problems are posed and solved. We don't know in advance which ideas and equations to ban, and, even if we did, it doesn't seem reasonable to expect that such a ban could be enforceable or effective.

To compound the difficulty still further, researchers making progress on general-purpose AI are often working on something else. As I have already argued, research on tool AI—those specific, innocuous applications such as game playing, medical diagnosis, and travel planning—often leads to progress on general-purpose techniques that are applicable to a wide range of other problems and move us closer to human-level AI.

For these reasons, it's very unlikely that the AI community—or the governments and corporations that control the laws and research budgets—will respond to the gorilla problem by ending progress in AI. If the gorilla problem can be solved only in this way, it isn't going to be solved.

The only approach that seems likely to work is to understand why it is that making better AI might be a bad thing. It turns out that we have known the answer for thousands of years.

The King Midas Problem

Norbert Wiener, whom we met in Chapter 1, had a profound impact on many fields, including artificial intelligence, cognitive science, and control theory. Unlike most of his contemporaries, he was particularly concerned with the unpredictability of complex systems operating in the real world. (He wrote his first paper on this topic at the age of ten.) He became convinced that the overconfidence of scientists and engineers in their ability to control their creations, whether military or civilian, could have disastrous consequences.

In 1950, Wiener published *The Human Use of Human Beings*,[5] whose front-cover blurb reads, "The 'mechanical brain' and similar machines can destroy human values or enable us to realize them as never before."[6] He gradually refined his ideas over time and by 1960 had identified one core issue: the impossibility of defining true human purposes correctly and completely. This, in turn, means that what I have called the standard model—whereby humans attempt to imbue machines with their own purposes—is destined to fail.

We might call this the *King Midas problem*: Midas, a legendary king in ancient Greek mythology, got exactly what he asked for— namely, that everything he touched should turn to gold. Too late, he discovered that this included his food, his drink, and his family members, and he died in misery and starvation. The same theme is ubiquitous in human mythology. Wiener cites Goethe's tale of the sorcerer's apprentice, who instructs the broom to fetch water—but doesn't say how much water and doesn't know how to make the broom stop.

A technical way of saying this is that we may suffer from a failure of *value alignment*—we may, perhaps inadvertently, imbue machines with objectives that are imperfectly aligned with our own. Until recently, we were shielded from the potentially catastrophic consequences by the limited capabilities of intelligent machines and the limited scope that they have to affect the world. (Indeed, most AI work was done with toy problems in research labs.) As Norbert Wiener put it in his 1964 book *God and Golem*,[7]

In the past, a partial and inadequate view of human purpose has been relatively innocuous only because it has been accompanied by technical limitations. . . . This is only one of the many places where human impotence has shielded us from the full destructive impact of human folly.

Unfortunately, this period of shielding is rapidly coming to an end.

We have already seen how content-selection algorithms on social media wrought havoc on society in the name of maximizing ad revenues. In case you are thinking to yourself that ad revenue maximization was already an ignoble goal that should never have been pursued, let's suppose instead that we ask some future superintelligent system to pursue the noble goal of finding a cure for cancer—ideally as quickly as possible, because someone dies from cancer every 3.5 seconds. Within hours, the AI system has read the entire biomedical literature and hypothesized millions of potentially effective but previously untested chemical compounds. Within weeks, it has induced multiple tumors of different kinds in every living human being so as to carry out medical trials of these compounds, this being the fastest way to find a cure. Oops.

If you prefer solving environmental problems, you might ask the machine to counter the rapid acidification of the oceans that results from higher carbon dioxide levels. The machine develops a new catalyst that facilitates an incredibly rapid chemical reaction between ocean and atmosphere and restores the oceans' pH levels. Unfortunately, a quarter of the oxygen in the atmosphere is used up in the process, leaving us to asphyxiate slowly and painfully. Oops.

These kinds of world-ending scenarios are unsubtle—as one might expect, perhaps, for world-ending scenarios. But there are many scenarios in which a kind of mental asphyxiation "steals upon us noiselessly and by imperceptible approaches." The prologue to Max Tegmark's *Life 3.0* describes in some detail a scenario in which a superintelligent machine gradually assumes economic and political control over the entire world while remaining essentially undetected. The Internet and the global-scale machines that it supports—the ones that already interact with billions of "users" on a daily basis—provide the perfect medium for the growth of machine control over humans.

I don't expect that the purpose put into such machines will be of the "take over the world" variety. It is more likely to be profit maximization or engagement maximization or, perhaps, even an apparently benign goal such as achieving higher scores on regular user happiness surveys or reducing our energy usage. Now, if we think of ourselves as entities whose actions are expected to achieve our objectives, there are two ways to change our behavior. The first is the old-fashioned way: leave our expectations and objectives unchanged, but change our circumstances—for example, by offering money, pointing a gun at us, or starving us into submission. That tends to be expensive and difficult for a computer to do. The second way is to change our expectations and objectives. This is much easier for a machine. It is in contact with you for hours every day, controls your access to information, and provides much of your entertainment through games, TV, movies, and social interaction.

The reinforcement learning algorithms that optimize social-media click-through have no capacity to reason about human behavior—in fact, they do not even know in any meaningful sense that humans exist. For machines with much greater understanding of human psychology, beliefs, and motivations, it should be relatively easy to gradually guide us in directions that increase the degree of satisfaction of the machine's objectives. For example, it might reduce our energy consumption by persuading us to have fewer children, eventually—and inadvertently—achieving the dreams of anti-natalist philosophers who wish to eliminate the noxious impact of humanity on the natural world.

With a bit of practice, you can learn to identify ways in which the achievement of more or less any fixed objective can result in arbitrarily bad outcomes. One of the most common patterns involves omitting something from the objective that you do actually care about. In such cases—as in the examples given above—the AI system will often find an optimal solution that sets the thing you do care about, but forgot to mention, to an extreme value. So, if you say to your self-driving car, "Take me to the airport as fast as possible!" and it interprets this literally, it will reach speeds of 180 miles per hour and you'll go to prison. (Fortunately, the self-driving cars currently contemplated won't accept such a request.) If you say, "Take me to the airport as fast as possible while not exceeding the speed limit," it will accelerate and brake as hard as possible, swerving in and out of traffic to maintain the maximum speed in between. It may even push other cars out of the way to gain a few seconds in the scrum at the airport terminal. And so on—eventually, you will add enough considerations so that the car's driving roughly approximates that of a skilled human driver taking someone to the airport in a bit of a hurry.

Driving is a simple task with only local impacts, and the AI systems currently being built for driving are not very intelligent. For these reasons, many of the potential failure modes can be anticipated; others will reveal themselves in driving simulators or in millions of miles of testing with professional drivers ready to take over if something goes wrong; still others will appear only later, when the cars are already on the road and something weird happens.

Unfortunately, with superintelligent systems that can have a global impact, there are no simulators and no do-overs. It's certainly very hard, and perhaps impossible, for mere humans to anticipate and rule out in advance all the disastrous ways the machine could choose to achieve a specified objective. Generally speaking, if you have one goal and a superintelligent machine has a different, conflicting goal, the machine gets what it wants and you don't.

Fear and Greed: Instrumental Goals

If a machine pursuing an incorrect objective sounds bad enough, there's worse. The solution suggested by Alan Turing—turning off the power at strategic moments—may not be available, for a very simple reason: *you can't fetch the coffee if you're dead.*

Let me explain. Suppose a machine has the objective of fetching the coffee. If it is sufficiently intelligent, it will certainly understand that it will fail in its objective if it is switched off before completing its mission. Thus, the objective of fetching coffee creates, as a necessary subgoal, the objective of disabling the off-switch. The same is true for curing cancer or calculating the digits of pi. There's really not a lot you can do once you're dead, so we can expect AI systems to act preemptively to preserve their own existence, given more or less *any* definite objective.

If that objective is in conflict with human preferences, then we have exactly the plot of *2001: A Space Odyssey*, in which the HAL 9000 computer kills four of the five astronauts on board the ship to prevent interference with its mission. Dave, the last remaining astronaut, manages to switch HAL off after an epic battle of wits— presumably to keep the plot interesting. But if HAL had been truly superintelligent, Dave would have been switched off.

It is important to understand that self-preservation doesn't have to be any sort of built-in instinct or prime directive in machines. (So Isaac Asimov's Third Law of Robotics,[8] which begins "A robot must protect its own existence," is completely unnecessary.) There is no need to build self-preservation in because it is an *instrumental goal*—a goal that is a useful subgoal of almost any original objective.[9] Any entity that has a definite objective will automatically act as if it also has instrumental goals.

In addition to being alive, having access to money is an instrumental goal within our current system. Thus, an intelligent machine might want money, not because it's greedy but because money is useful for achieving all sorts of goals. In the movie *Transcendence*, when Johnny Depp's brain is uploaded into the quantum supercomputer, the first thing the machine does is copy itself onto millions of other computers on the Internet so that it cannot be switched off. The second thing it does is make a quick killing on the stock market to fund its expansion plans.

And what, exactly, are those expansion plans? They include designing and building a much larger quantum supercomputer; doing AI research; and discovering new knowledge of physics, neuroscience, and biology. These resource objectives—computing power, algorithms, and knowledge—are also instrumental goals, useful for achieving any overarching objective.[10] They seem harmless enough until one realizes that the acquisition process will continue without limit. This seems to create inevitable conflict with humans. And of course, the machine, equipped with ever-better models of human decision making, will anticipate and defeat our every move in this conflict.

Intelligence Explosions

I. J. Good was a brilliant mathematician who worked with Alan Turing at Bletchley Park, breaking German codes during World War II. He shared Turing's interests in machine intelligence and statistical inference. In 1965, he wrote what is now his best-known paper, "Speculations Concerning the First Ultraintelligent Machine."[11] The first sentence suggests that Good, alarmed by the nuclear brinkmanship of the Cold War, regarded AI as a possible savior for humanity: "The survival of man depends on the early construction of an ultraintelligent machine." As the paper proceeds, however, he becomes more circumspect. He introduces the notion of an *intelligence explosion*, but, like Butler, Turing, and Wiener before him, he worries about losing control:

> Let an ultraintelligent machine be defined as a machine that can far surpass all the intellectual activities of any man however clever. Since the design of machines is one of these intellectual activities, an ultraintelligent machine could design even better machines; there would then unquestionably be an "intelligence explosion,"

and the intelligence of man would be left far behind. Thus the first ultraintelligent machine is the last invention that man need ever make, provided that the machine is docile enough to tell us how to keep it under control. It is curious that this point is made so seldom outside science fiction.

This paragraph is a staple of any discussion of superintelligent AI, although the caveats at the end are usually left out. Good's point can be strengthened by noting that not only *could* the ultraintelligent machine improve its own design; it's likely that it *would* do so because, as we have seen, an intelligent machine expects to benefit from improving its hardware and software. The possibility of an intelligence explosion is often cited as the main source of risk to humanity from AI because it would give us so little time to solve the control problem.[12]

Good's argument certainly has plausibility via the natural analogy to a chemical explosion in which each molecular reaction releases enough energy to initiate more than one additional reaction. On the other hand, it is logically possible that there are diminishing returns to intelligence improvements, so that the process peters out rather than exploding.[13] There's no obvious way to prove that an explosion will *necessarily* occur.

The diminishing-returns scenario is interesting in its own right. It could arise if it turns out that achieving a given percentage improvement becomes much harder as the machine becomes more intelligent. (I'm assuming for the sake of argument that general-purpose machine intelligence is measurable on some kind of linear scale, which I doubt will ever be strictly true.) In that case, humans won't be able to create superintelligence either. If a machine that is already superhuman runs out of steam when trying to improve its own intelligence, then humans will run out of steam even sooner.

Now, I've never heard a serious argument to the effect that creating any given level of machine intelligence is simply beyond the capacity of human ingenuity, but I suppose one must concede it's logically possible. "Logically possible" and "I'm willing to bet the future of the human race on it" are, of course, two completely different things. Betting against human ingenuity seems like a losing strategy.

If an intelligence explosion does occur, and if we have not already solved the problem of controlling machines with only slightly superhuman intelligence—for example, if we cannot prevent them from making these recursive self-improvements—then we would have no time left to solve the control problem and the game would be over. This is Bostrom's *hard takeoff* scenario, in which the machine's intelligence increases astronomically in just days or weeks. In Turing's words, it is "certainly something which can give us anxiety."

The possible responses to this anxiety seem to be to retreat from AI research, to deny that there are risks inherent in developing advanced AI, to understand and mitigate the risks through the design of AI systems that necessarily remain under human control, and to resign—simply to cede the future to intelligent machines.

Denial and mitigation are the subjects of the remainder of the book. As I have already argued, retreat from AI research is both unlikely to happen (because the benefits forgone are too great) and very difficult to bring about. Resignation seems to be the worst possible response. It is often accompanied by the idea that AI systems that are more intelligent than us somehow *deserve* to inherit the planet, leaving humans to go gentle into that good night, comforted by the thought that our brilliant electronic progeny are busy pursuing their objectives. This view was promulgated by the roboticist and futurist Hans Moravec,[14] who writes, "The immensities of cyberspace will be teeming with unhuman superminds, engaged in affairs that are to human concerns as ours are to those of bacteria." This seems to be a mistake. Value, for humans, is defined primarily by conscious human experience. If there are no humans and no other conscious entities whose subjective experience matters to us, there is nothing of value occurring.

6

THE NOT-SO-GREAT
AI DEBATE

"The implications of introducing a second intelligent species onto Earth are far-reaching enough to deserve hard thinking."[1] So ended *The Economist* magazine's review of Nick Bostrom's *Superintelligence*. Most would interpret this as a classic example of British understatement. Surely, you might think, the great minds of today are already doing this hard thinking—engaging in serious debate, weighing up the risks and benefits, seeking solutions, ferreting out loopholes in solutions, and so on. Not yet, as far as I am aware.

When one first introduces these ideas to a technical audience, one can see the thought bubbles popping out of their heads, beginning with the words "But, but, but . . ." and ending with exclamation marks.

The first kind of *but* takes the form of denial. The deniers say, "But this can't be a real problem, because XYZ." Some of the XYZs reflect a reasoning process that might charitably be described as wishful thinking, while others are more substantial. The second kind of *but* takes the form of deflection: accepting that the problems are real but arguing that we shouldn't try to solve them, either because they're unsolvable or because there are more important things to focus on than the end of civilization or because it's best not to mention them at all. The third kind of *but* takes the form of an oversimplified, instant solution: "But can't we just do ABC?" As with denial, some of the ABCs are instantly regrettable. Others, perhaps by accident, come closer to identifying the true nature of the problem.

I don't mean to suggest that there cannot be any reasonable objections to the view that poorly designed superintelligent machines would present a serious risk to humanity. It's just that I have yet to see such an objection. Since the issue seems to be so important, it deserves a public debate of the highest quality. So, in the interests of having that debate, and in the hope that the reader will contribute to it, let me provide a quick tour of the highlights so far, such as they are.

Denial

Denying that the problem exists at all is the easiest way out. Scott Alexander, author of the *Slate Star Codex* blog, began a well-known article on AI risk as follows:[2] "I first became interested in AI risk back around 2007. At the time, most people's response to the topic was 'Haha, come back when anyone believes this besides random Internet crackpots.'"

Instantly regrettable remarks

A perceived threat to one's lifelong vocation can lead a perfectly intelligent and usually thoughtful person to say things they might wish to retract on further analysis. That being the case, I will not name the authors of the following arguments, all of whom are well-known AI researchers. I've included refutations of the arguments, even though they are quite unnecessary.

- Electronic calculators are superhuman at arithmetic. Calculators didn't take over the world; therefore, there is no reason to worry about superhuman AI.
  - *Refutation: intelligence is not the same as arithmetic, and the arithmetic ability of calculators does not equip them to take over the world.*
- Horses have superhuman strength, and we don't worry about proving that horses are safe; so we needn't worry about proving that AI systems are safe.
  - *Refutation: intelligence is not the same as physical strength, and the strength of horses does not equip them to take over the world.*
- Historically, there are zero examples of machines killing millions of humans, so, by induction, it cannot happen in the future.
  - *Refutation: there's a first time for everything, before which there were zero examples of it happening.*
- No physical quantity in the universe can be infinite, and that includes intelligence, so concerns about superintelligence are overblown.
  - *Refutation: superintelligence doesn't need to be infinite to be problematic; and physics allows computing devices billions of times more powerful than the human brain.*
- We don't worry about species-ending but highly unlikely possibilities such as black holes materializing in near-Earth orbit, so why worry about superintelligent AI?
  - *Refutation: if most physicists on Earth were working to make such black holes, wouldn't we ask them if it was safe?*

It's complicated

It is a staple of modern psychology that a single IQ number cannot characterize the full richness of human intelligence.[3] There are, the theory says, different dimensions of intelligence: spatial, logical, linguistic, social, and so on. Alice, our soccer player from Chapter 2, might have more spatial intelligence than her friend Bob, but less social intelligence. Thus, we cannot line up all humans in strict order of intelligence.

This is even more true of machines, because their abilities are much narrower. The Google search engine and AlphaGo have almost nothing in common, besides being products of two subsidiaries of the same parent corporation, and so it makes no sense to say that one is more intelligent than the other. This makes notions of "machine IQ" problematic and suggests that it's misleading to describe the future as a one-dimensional IQ race between humans and machines.

Kevin Kelly, founding editor of *Wired* magazine and a remarkably perceptive technology commentator, takes this argument one step further. In "The Myth of a Superhuman AI,"[4] he writes, "Intelligence is not a single dimension, so 'smarter than humans' is a meaningless concept." In a single stroke, all concerns about superintelligence are wiped away.

Now, one obvious response is that a machine could exceed human capabilities in *all* relevant dimensions of intelligence. In that case, even by Kelly's strict standards, the machine would be smarter than a human. But this rather strong assumption is not necessary to refute Kelly's argument. Consider the chimpanzee. Chimpanzees probably have better short-term memory than humans, even on human-oriented tasks such as recalling sequences of digits.[5] Short-term memory is an important dimension of intelligence. By Kelly's argument, then, humans are not smarter than chimpanzees; indeed, he would claim that "smarter than a chimpanzee" is a meaningless concept. This is cold comfort to the chimpanzees (and bonobos, gorillas, orangutans, whales, dolphins, and so on) whose species survive only because we deign to allow it. It is colder comfort still to all those species that we have already wiped out. It's also cold comfort to humans who might be worried about being wiped out by machines.

It's impossible

Even before the birth of AI in 1956, august intellectuals were har-rumphing and saying that intelligent machines were impossible. Alan Turing devoted much of his seminal 1950 paper, "Computing Machinery and Intelligence," to refuting these arguments. Ever since, the AI community has been fending off similar claims of impossibility from philosophers,[6] mathematicians,[7] and others. In the current debate over superintelligence, several philosophers have exhumed these impossibility claims to prove that humanity has nothing to fear.[8,9] This comes as no surprise.

The One Hundred Year Study on Artificial Intelligence, or AI100, is an ambitious, long-term project housed at Stanford University. Its goal is to keep track of AI, or, more precisely, to "study and anticipate how the effects of artificial intelligence will ripple through every aspect of how people work, live and play." Its first major report, "Artificial Intelligence and Life in 2030," *does* come as a surprise.[10] As might be expected, it emphasizes the benefits of AI in areas such as medical diagnosis and automotive safety. What's unexpected is the claim that "unlike in the movies, there is no race of superhuman robots on the horizon or probably even possible."

To my knowledge, this is the first time that serious AI researchers have publicly espoused the view that human-level or superhuman AI is impossible—and this in the middle of a period of extremely rapid progress in AI research, when barrier after barrier is being breached. It's as if a group of leading cancer biologists announced that they had been fooling us all along: they've always known that there will never be a cure for cancer.

What could have motivated such a volte-face? The report provides no arguments or evidence whatever. (Indeed, what evidence could there be that no physically possible arrangement of atoms outperforms the human brain?) I suspect there are two reasons. The first is the natural desire to disprove the existence of the gorilla problem, which presents a very uncomfortable prospect for the AI researcher; certainly, if human-level AI is impossible, the gorilla problem is neatly dispatched. The second reason is *tribalism*—the instinct to circle the wagons against what are perceived to be "attacks" on AI.

It seems odd to perceive the claim that superintelligent AI is possible as an attack on AI, and even odder to defend AI by saying that AI will never succeed in its goals. We cannot insure against future catastrophe simply by betting against human ingenuity.

We have made such bets before and lost. As we saw earlier, the physics establishment of the early 1930s, personified by Lord Rutherford, confidently believed that extracting atomic energy was impossible; yet Leo Szilard's invention of the neutron-induced nuclear chain reaction in 1933 proved that confidence to be misplaced.

Szilard's breakthrough came at an unfortunate time: the beginning of an arms race with Nazi Germany. There was no possibility of developing nuclear technology for the greater good. A few years later, having demonstrated a nuclear chain reaction in his laboratory, Szilard wrote, "We switched everything off and went home. That night, there was very little doubt in my mind that the world was headed for grief."

It's too soon to worry about it

It's common to see sober-minded people seeking to assuage public concerns by pointing out that because human-level AI is not likely to arrive for several decades, there is nothing to worry about. For example, the AI100 report says there is "no cause for concern that AI is an imminent threat to humankind."

This argument fails on two counts. The first is that it attacks a straw man. The reasons for concern are *not* predicated on imminence. For example, Nick Bostrom writes in *Superintelligence*, "It is no part of the argument in this book that we are on the threshold of a big breakthrough in artificial intelligence, or that we can predict with any precision when such a development might occur." The second is that a long-term risk can still be cause for immediate concern. The right time to worry about a potentially serious problem for humanity depends not just on when the problem will occur but also on how long it will take to prepare and implement a solution.

For example, if we were to detect a large asteroid on course to collide with Earth in 2069, would we say it's too soon to worry? Quite the opposite! There would be a worldwide emergency project to develop the means to counter the threat. We wouldn't wait until 2068 to start working on a solution, because we can't say in advance how much time is needed. Indeed, NASA's Planetary Defense project is *already* working on possible solutions, even though "no known asteroid poses a significant risk of impact with Earth over the next 100 years." In case that makes you feel complacent, they also say, "About 74 percent of near-Earth objects larger than 460 feet still remain to be discovered."

And if we consider the global catastrophic risks from climate change, which are predicted to occur later in this century, is it too soon to take action to prevent them? On the contrary, it may be too late. The relevant time scale for superhuman AI is less predictable, but of course that means it, like nuclear fission, might arrive considerably sooner than expected.

One formulation of the "it's too soon to worry" argument that has gained currency is Andrew Ng's assertion that "it's like worrying about overpopulation on Mars."[11] (He later upgraded this from Mars to Alpha Centauri.) Ng, a former Stanford professor, is a leading expert on machine learning, and his views carry some weight. The assertion appeals to a convenient analogy: not only is the risk easily managed and far in the future but also it's extremely unlikely we'd even try to move billions of humans to Mars in the first place. The analogy is a false one, however. We are *already* devoting huge scientific and technical resources to creating ever-more-capable AI systems, with very little thought devoted to what happens if we succeed. A more apt analogy, then, would be working on a plan to move the human race to

Mars with no consideration for what we might breathe, drink, or eat once we arrive. Some might call this plan unwise. Alternatively, one could take Ng's point literally, and respond that landing even a single person on Mars would constitute overpopulation, because Mars has a carrying capacity of zero. Thus, groups that are currently planning to send a handful of humans to Mars *are* worrying about overpopulation on Mars, which is why they are developing life-support systems.

We're the experts

In every discussion of technological risk, the pro-technology camp wheels out the claim that all concerns about risk arise from ignorance. For example, here's Oren Etzioni, CEO of the Allen Institute for AI and a noted researcher in machine learning and natural language understanding:[12]

> At the rise of every technology innovation, people have been scared. From the weavers throwing their shoes in the mechanical looms at the beginning of the industrial era to today's fear of killer robots, our response has been driven by not knowing what impact the new technology will have on our sense of self and our livelihoods. And when we don't know, our fearful minds fill in the details.

Popular Science published an article titled "Bill Gates Fears AI, but AI Researchers Know Better":[13]

> When you talk to A.I. researchers—again, genuine A.I. researchers, people who grapple with making systems that work at all, much less work too well—they are not worried about superintelligence sneaking up on them, now or in the future. Contrary to the spooky stories that Musk seems intent on telling, A.I. researchers aren't frantically installing firewalled summoning chambers and self-destruct countdowns.

This analysis was based on a sample of four, all of whom in fact said in their interviews that the long-term safety of AI was an important issue.

Using very similar language to the *Popular Science* article, David Kenny, at that time a vice president at IBM, wrote a letter to the US Congress that included the following reassuring words:[14]

> When you actually do the science of machine intelligence, and when you actually apply it in the real world of business and society—as we have done at IBM to create our pioneering cognitive computing system, Watson—you understand that this technology does not support the fear-mongering commonly associated with the AI debate today.

The message is the same in all three cases: "Don't listen to them; we're the experts." Now, one can point out that this is really an ad hominem argument that attempts to refute the message by delegitimizing the messengers, but even if one takes it at face value, the argument doesn't hold water. Elon Musk, Stephen Hawking, and Bill Gates are certainly very familiar with scientific and technological reasoning, and Musk and Gates in particular have supervised and invested in many AI research projects. And it would be even less plausible to argue that Alan Turing, I. J. Good, Norbert Wiener, and Marvin Minsky are unqualified to discuss AI. Finally, Scott Alexander's blog piece mentioned earlier, which is titled "AI Researchers on AI Risk," notes that "AI researchers, including some of the leaders in the field, have been instrumental in raising issues about AI risk and superintelligence from the very beginning." He lists several such researchers, and the list is now much longer.

Another standard rhetorical move for the "defenders of AI" is to describe their opponents as Luddites. Oren Etzioni's reference to "weavers throwing their shoes in the mechanical looms" is just this: the Luddites were artisan weavers in the early nineteenth century protesting the introduction of machinery to replace their skilled labor. In 2015, the Information Technology and Innovation Foundation gave its annual Luddite Award to "alarmists touting an artificial intelligence apocalypse." It's an odd definition of "Luddite" that includes Turing, Wiener, Minsky, Musk, and Gates, who rank among the most prominent contributors to technological progress in the twentieth and twenty-first centuries.

The accusation of Luddism represents a misunderstanding of the nature of the concerns raised and the purpose for raising them. It is as if one were to accuse nuclear engineers of Luddism if they point out the need for control of the fission reaction. As with the strange phenomenon of AI researchers suddenly claiming that AI is impossible, I think we can attribute this puzzling episode to tribalism in defense of technological progress.

Deflection

Some commentators are willing to accept that the risks are real, but still present arguments for doing nothing. These arguments include the impossibility of doing anything, the importance of doing something else entirely, and the need to keep quiet about the risks.

You can't control research

A common answer to suggestions that advanced AI might present risks to humanity is to claim that banning AI research is impossible. Note the mental leap here: "Hmm, someone is discussing risks! They must be proposing a ban on my research!!" This mental leap might be appropriate in a discussion of risks based only on the gorilla problem, and I would tend to agree that solving the gorilla problem by preventing the creation of superintelligent AI would require some kind of constraints on AI research.

Recent discussions of risks have, however, focused not on the general gorilla problem (journalistically speaking, the humans vs. super-intelligence smackdown) but on the King Midas problem and variants thereof. Solving the King Midas problem also solves the gorilla problem—not by preventing superintelligent AI or finding a way to defeat it but by ensuring that it is never in conflict with humans in the first place. Discussions of the King Midas problem generally avoid proposing that AI research be curtailed; they merely suggest that attention be paid to the issue of preventing negative consequences of poorly designed systems. In the same vein, a discussion of the risks of containment failure in nuclear plants should be interpreted not as an attempt to ban nuclear physics research but as a suggestion to focus more effort on solving the containment problem.

There is, as it happens, a very interesting historical precedent for cutting off research. In the early 1970s, biologists began to be concerned that novel recombinant DNA methods—splicing genes from one organism into another—might create substantial risks for human health and the global ecosystem. Two meetings at Asilomar in California in 1973 and 1975 led first to a moratorium on such experiments and then to detailed biosafety guidelines consonant with the risks posed by any proposed experiment.[15] Some classes of experiments, such as those involving toxin genes, were deemed too hazardous to be allowed.

Immediately after the 1975 meeting, the National Institutes of Health (NIH), which funds virtually all basic medical research in the United States, began the process of setting up the Recombinant DNA Advisory Committee. The RAC, as it is known, was instrumental in developing the NIH guidelines that essentially implemented the Asilomar recommendations. Since 2000, those guidelines have included a ban on funding approval for any protocol involving *human germline alteration*—the modification of the human genome in ways that can be inherited by subsequent generations. This ban was followed by legal prohibitions in over fifty countries.

The goal of "improving the human stock" had been one of the

dreams of the eugenics movement in the late nineteenth and early twentieth centuries. The development of CRISPR-Cas9, a very precise method for genome editing, has reignited this dream. An international summit held in 2015 left the door open for future applications, calling for restraint until "there is broad societal consensus about the appropriateness of the proposed application."[16] In November 2018, the Chinese scientist He Jiankui announced that he had edited the genomes of three human embryos, at least two of which had led to live births. An international outcry followed, and at the time of writing, Jiankui appears to be under house arrest. In March 2019, an international panel of leading scientists called explicitly for a formal moratorium.[17]

The lesson of this debate for AI is mixed. On the one hand, it shows that we *can* refrain from proceeding with an area of research that has huge potential. The international consensus against germline alteration has been almost completely successful up to now. The fear that a ban would simply drive the research underground, or into countries with no regulation, has not materialized. On the other hand, germline alteration is an easily identifiable process, a specific use case of more general knowledge about genetics that requires specialized equipment and real humans to experiment on. Moreover, it falls within an area—reproductive medicine—that is already subject to close oversight and regulation. These characteristics do not apply to general-purpose AI, and, as yet, no one has come up with any plausible form that a regulation to curtail AI research might take.

Whataboutery

I was introduced to the term *whataboutery* by an adviser to a British politician who had to deal with it on a regular basis at public meetings. No matter the topic of the speech he was giving, someone would invariably ask, "What about the plight of the Palestinians?"

In response to any mention of risks from advanced AI, one is likely to hear, "What about the benefits of AI?" For example, here is Oren Etzioni:[18]

> Doom-and-gloom predictions often fail to consider the potential benefits of AI in preventing medical errors, reducing car accidents, and more.

And here is Mark Zuckerberg, CEO of Facebook, in a recent media-fueled exchange with Elon Musk:[19]

> If you're arguing against AI, then you're arguing against safer cars that aren't going to have accidents and you're arguing against being able to better diagnose people when they're sick.

Leaving aside the tribal notion that anyone mentioning risks is "against AI," both Zuckerberg and Etzioni are arguing that to talk about risks is to ignore the potential benefits of AI or even to negate them.

This is precisely backwards, for two reasons. First, if there were no potential benefits of AI, there would be no economic or social impetus for AI research and hence no danger of ever achieving human-level AI. We simply wouldn't be having this discussion at all. Second, *if the risks are not successfully mitigated, there will be no benefits.* The potential benefits of nuclear power have been greatly reduced because of the partial core meltdown at Three Mile Island in 1979, the uncontrolled reaction and catastrophic releases at Chernobyl in 1986, and the multiple meltdowns at Fukushima in 2011. Those disasters severely curtailed the growth of the nuclear industry. Italy abandoned nuclear power in 1990 and Belgium, Germany, Spain, and Switzerland have announced plans to do so. Since 1990, the worldwide rate of commissioning of nuclear plants has been about a tenth of what it was before Chernobyl.

Silence

The most extreme form of deflection is simply to suggest that we should keep silent about the risks. For example, the aforementioned AI100 report includes the following admonition:

> If society approaches these technologies primarily with fear and suspicion, missteps that slow AI's development or drive it underground will result, impeding important work on ensuring the safety and reliability of AI technologies.

Robert Atkinson, director of the Information Technology and Innovation Foundation (the very same foundation that gives out the Luddite Award), made a similar argument in a 2015 debate.[20] While there are valid questions about precisely how risks should be described when talking to the media, the overall message is clear: "Don't mention the risks; it would be bad for funding." Of course, if no one were aware of the risks, there would be no funding for research on risk mitigation and no reason for anyone to work on it.

The renowned cognitive scientist Steven Pinker gives a more optimistic version of Atkinson's argument. In his view, the "culture of safety in advanced societies" will ensure that all serious risks from AI will be eliminated; therefore, it is inappropriate and counterproductive to call attention to those risks.[21] Even if we disregard the fact that our advanced culture of safety has led to Chernobyl, Fukushima, and runaway global warming, Pinker's argument entirely misses the point. The culture of safety consists precisely of people pointing to possible failure modes and finding ways to ensure they don't happen. (And with AI, the standard model *is* the failure mode.) Saying that it's ridiculous to point to a failure mode because the culture of safety will fix it anyway is like saying no one should call an ambulance when they see a hit-and-run accident because someone will call an ambulance.

In attempting to portray the risks to the public and to policy makers, AI researchers are at a disadvantage compared to nuclear physicists. The physicists did not need to write books explaining to the public that assembling a critical mass of highly enriched uranium might present a risk, because the consequences had already been demonstrated at Hiroshima and Nagasaki. It did not require a great deal of further persuasion to convince governments and funding agencies that safety was important in developing nuclear energy.

Tribalism

In Butler's *Erewhon*, focusing on the gorilla problem leads to a premature and false dichotomy between pro-machinists and anti-machinists. The pro-machinists believe the risk of machine domination to be minimal or nonexistent; the anti-machinists believe it to be insuperable unless all machines are destroyed. The debate becomes tribal, and no one tries to solve the underlying problem of retaining human control over the machines.

To varying degrees, all the major technological issues of the twentieth century—nuclear power, genetically modified organisms (GMOs), and fossil fuels—succumbed to tribalism. On each issue, there are two sides, pro and anti. The dynamics and outcomes of each have been different, but the symptoms of tribalism are similar: mutual distrust and denigration, irrational arguments, and a refusal to concede any (reasonable) point that might favor the other tribe. On the pro-technology side, one sees denial and concealment of risks combined with accusations of Luddism; on the anti side, one sees a conviction that the risks are insuperable and the problems unsolvable. A member of the pro-technology tribe who is too honest about a problem is viewed as a traitor, which is particularly unfortunate as the pro-technology tribe usually includes most of the people qualified to solve the problem. A member of the anti-technology tribe who discusses possible mitigations is also a traitor, because it is the technology itself that has come to be viewed as evil, rather than its possible effects. In this way, only the most extreme voices—those least likely to be listened to by the other side—can speak for each tribe.

In 2016, I was invited to No. 10 Downing Street to meet with some of then prime minister David Cameron's advisers. They were worried that the AI debate was starting to resemble the GMO debate—which, in Europe, had led to what the advisers considered to be premature and overly restrictive regulations on GMO production and labeling. They wanted to avoid the same thing happening to AI. Their concerns had some validity: the AI debate *is* in danger of becoming tribal, of creating pro-AI and anti-AI camps. This would be damaging to the field because it's simply not true that being concerned about the risks inherent in advanced AI is an anti-AI stance. A physicist who is concerned about the risks of nuclear war or the risk of a poorly designed nuclear reactor exploding is not "anti-physics." To say that AI will be powerful enough to have a global impact is a compliment to the field rather than an insult.

It is essential that the AI community own the risks and work to mitigate them. The risks, to the extent that we understand them, are neither minimal nor insuperable. We need to do a substantial amount of work to avoid them, including reshaping and rebuilding the foundations of AI.

Can't We Just . . .

. . . switch it off?

Once they understand the basic idea of existential risk, whether in the form of the gorilla problem or the King Midas problem, many people—myself included—immediately begin casting around for an easy solution. Often, the first thing that comes to mind is switching off the machine. For example, Alan Turing himself, as quoted earlier, speculates that we might "keep the machines in a subservient posi-tion, for instance by turning off the power at strategic moments."

This won't work, for the simple reason that a superintelligent entity will *already have thought of that possibility* and taken steps to prevent it. And it will do that not because it wants to stay alive but because it is pursuing whatever objective we gave it and knows that it will fail if it is switched off.

There are some systems being contemplated that really cannot be switched off without ripping out a lot of the plumbing of our civilization. These are systems implemented as so-called smart contracts in the blockchain. The *blockchain* is a highly distributed form of computing and record keeping based on encryption; it is specifically designed so that no datum can be deleted and no smart contract can be interrupted without essentially taking control of a very large number of machines and undoing the chain, which might in turn destroy a large part of the Internet and/or the financial system. It is debatable whether this incredible robustness is a feature or a bug. It's certainly a tool that a superintelligent AI system could use to protect itself.

. . . put it in a box?

If you can't switch AI systems off, can you seal the machines inside a kind of firewall, extracting useful question-answering work from them but never allowing them to affect the real world directly? This is the idea behind Oracle AI, which has been discussed at length in the AI safety community.[22] An Oracle AI system can be arbitrarily intelligent, but can answer only yes or no (or give corresponding probabilities) to each question. It can access all the information the human race possesses through a read-only connection—that is, it has no direct access to the Internet. Of course, this means giving up on superintelligent robots, assistants, and many other kinds of AI systems, but a trustworthy Oracle AI would still have enormous economic value because we could ask it questions whose answers are important to us, such as whether Alzheimer's disease is caused by an infectious organism or whether it's a good idea to ban autonomous weapons. Thus, the Oracle AI is certainly an interesting possibility.

Unfortunately, there are some serious difficulties. First, the Oracle AI system will be at least as assiduous in understanding the physics and origins of its world—the computing resources, their mode of operation, and the mysterious entities that produced its information store and are now asking questions—as we are in understanding ours. Second, if the objective of the Oracle AI system is to provide accurate answers to questions in a reasonable amount of time, it will have an incentive to break out of its cage to acquire more computational resources and to control the questioners so that they ask only simple questions. And, finally, we have yet to invent a firewall that is secure against ordinary humans, let alone superintelligent machines.

I think there *might* be solutions to some of these problems, particularly if we limit Oracle AI systems to be provably sound logical or Bayesian calculators. That is, we could insist that the algorithm can output only a conclusion that is warranted by the information provided, and we could check mathematically that the algorithm satisfies this condition. This still leaves the problem of controlling the process that decides *which* logical or Bayesian computations to do, in order to reach the strongest possible conclusion as quickly as possible. Because this process has an incentive to reason quickly, it has an incentive to acquire computational resources and of course to preserve its own existence.

In 2018, the Center for Human-Compatible AI at Berkeley ran a workshop at which we asked the question, "What would you do if you knew for certain that superintelligent AI would be achieved within a decade?" My answer was as follows: persuade the developers to hold off on building a general-purpose intelligent agent—one that can choose its own actions in the real world—and build an Oracle AI instead. Meanwhile, we would work on solving the problem of making Oracle AI systems provably safe to the extent possible. The reason this strategy might work is twofold: first, a superintelligent Oracle AI system would still be worth trillions of dollars, so the developers might be willing to accept this restriction; and second, controlling Oracle AI systems is almost certainly easier than controlling a general-purpose intelligent agent, so we'd have a better chance of solving the problem within the decade.

. . . work in human–machine teams?

A common refrain in the corporate world is that AI is no threat to employment or to humanity because we'll just have collaborative human–AI teams. For example, David Kenny's letter to Congress, quoted earlier in this chapter, stated that "high-value artificial intelligence systems are specifically designed to augment human intelligence, not replace workers."[23]

While a cynic might suggest that this is merely a public relations ploy to sugarcoat the process of eliminating human employees from the corporations' clients, I think it does move the ball forward a few inches. Collaborative human–AI teams are indeed a desirable goal. Clearly, a team will be unsuccessful if the objectives of the team members are not aligned, so the emphasis on human–AI teams highlights the need to solve the core problem of value alignment. Of course, highlighting the problem is not the same as solving it.

. . . merge with the machines?

Human–machine teaming, taken to its extreme, becomes a human–machine merger in which electronic hardware is attached directly to the brain and forms part of a single, extended, conscious entity. The futurist Ray Kurzweil describes the possibility as follows:[24]

We are going to directly merge with it, we are going to become the AIs. . . . As you get to the late 2030s or 2040s, our thinking will be predominately non-biological and the non-biological part will ultimately be so intelligent and have such vast capacity it'll be able to model, simulate and understand fully the biological part.

Kurzweil views these developments in a positive light. Elon Musk, on the other hand, views the human–machine merger primarily as a defensive strategy:[25]

> If we achieve tight symbiosis, the AI wouldn't be "other"—it would be you and [it would have] a relationship to your cortex analogous to the relationship your cortex has with your limbic system. . . . We're going to have the choice of either being left behind and being effectively useless or like a pet—you know, like a house cat or something—or eventually figuring out some way to be symbiotic and merge with AI.

Musk's Neuralink Corporation is working on a device dubbed "neural lace" after a technology described in Iain Banks's Culture novels. The aim is to create a robust, permanent connection between the human cortex and external computing systems and networks. There are two main technical obstacles: first, the difficulties of connecting an electronic device to brain tissue, supplying it with power, and connecting it to the outside world; and second, the fact that we understand almost nothing about the neural implementation of higher levels of cognition in the brain, so we don't know where to connect the device and what processing it should do.

I am not completely convinced that the obstacles in the preceding paragraph are insuperable. First, technologies such as *neural dust* are rapidly reducing the size and power requirements of electronic devices that can be attached to neurons and provide sensing, stimulation, and transcranial communication.[26] (The technology as of 2018 had reached a size of about one cubic millimeter, so *neural grit* might be a more accurate term.) Second, the brain itself has remarkable powers of adaptation. It used to be thought, for example, that we would have to understand the code that the brain uses to control the arm muscles before we could connect a brain to a robot arm success-fully, and that we would have to understand the way the cochlea ana-lyzes sound before we could build a replacement for it. It turns out, instead, that the brain does most of the work for us. It quickly learns how to make the robot arm do what its owner wants, and how to map the output of a cochlear implant to intelligible sounds. It's entirely possible that we may hit upon ways to provide the brain with addi-tional memory, with communication channels to computers, and per-haps even with communication channels to other brains—all without ever really understanding how any of it works.[27]

Regardless of the technological feasibility of these ideas, one has to ask whether this direction represents the best possible future for hu-manity. If humans need brain surgery merely to survive the threat posed by their own technology, perhaps we've made a mistake some-where along the line.

. . . avoid putting in human goals?

A common line of reasoning has it that problematic AI behaviors arise from putting in specific *kinds* of objectives; if these are left out, everything will be fine. Thus, for example, Yann LeCun, a pioneer of deep learning and director of AI research at Facebook, often cites this idea when downplaying the risk from AI:[28]

> There is no reason for AIs to have self-preservation instincts, jeal-ousy, etc. . . . AIs will not have these destructive "emotions" unless we build these emotions into them. I don't see why we would want to do that.

In a similar vein, Steven Pinker provides a gender-based analysis:[29]

AI dystopias project a parochial alpha-male psychology onto the concept of intelligence. They assume that superhumanly intelligent robots would develop goals like deposing their masters or taking over the world. . . . It's telling that many of our techno-prophets don't entertain the possibility that artificial intelligence will naturally develop along female lines: fully capable of solving problems, but with no desire to annihilate innocents or dominate the civilization.

As we have already seen in the discussion of instrumental goals, it doesn't matter whether we build in "emotions" or "desires" such as self-preservation, resource acquisition, knowledge discovery, or, in the extreme case, taking over the world. The machine is going to have those emotions anyway, as subgoals of any objective we do build in—and regardless of its gender. For a machine, death isn't bad per se. Death is to be avoided, nonetheless, because it's hard to fetch the coffee if you're dead.

An even more extreme solution is to avoid putting objectives into the machine altogether. Voilà, problem solved. Alas, it's not as simple as that. Without objectives, there is no intelligence: any action is as good as any other, and the machine may as well be a random number generator. Without objectives, there is also no reason for the machine to prefer a human paradise to a planet turned into a sea of paperclips (a scenario described at length by Nick Bostrom). Indeed, the latter outcome may be utopian for the iron-eating bacterium *Thiobacillus ferrooxidans*. Absent some notion that human preferences matter, who is to say the bacterium is wrong?

A common variant on the "avoid putting in objectives" idea is the notion that a sufficiently intelligent system will necessarily, as a consequence of its intelligence, develop the "right" goals on its own. Often, proponents of this notion appeal to the theory that people of greater intelligence tend to have more altruistic and lofty objectives—a view that may be related to the self-conception of the proponents.

The idea that it is possible to perceive objectives in the world was discussed at length by the famous eighteenth-century philosopher David Hume in *A Treatise of Human Nature*.[30] He called it the *is-ought problem* and concluded that it was simply a mistake to think that moral imperatives could be deduced from natural facts. To see why, consider, for example, the design of a chessboard and chess pieces. One cannot perceive in these the goal of checkmate, for the same chessboard and pieces can be used for suicide chess or indeed many other games still to be invented.

Nick Bostrom, in *Superintelligence*, presents the same underlying idea in a different form, which he calls the *orthogonality thesis*:

> Intelligence and final goals are orthogonal: more or less any level of intelligence could in principle be combined with more or less any final goal.

Here, *orthogonal* means "at right angles" in the sense that the degree of intelligence is one axis defining an intelligent system and its goals are another axis, and we can vary these independently. For example, a self-driving car can be given any particular address as its destination; making the car a better driver doesn't mean that it will start refusing to go to addresses that are divisible by seventeen. By the same token, it is easy to imagine that a general-purpose intelligent system could be given more or less any objective to pursue—including maximizing the number of paperclips or the number of known digits of pi. This is just how reinforcement learning systems and other kinds of reward optimizers work: the algorithms are completely general and accept *any* reward signal. For engineers and computer scientists operating within the standard model, the orthogonality thesis is just a given.

The idea that intelligent systems could simply observe the world to acquire the goals that should be pursued suggests that a sufficiently intelligent system will naturally abandon its initial objective in favor of the "right" objective. It's hard to see why a rational agent would do this. Furthermore, it presupposes that there is a "right" objective out there in the world; it would have to be an objective on which iron-eating bacteria and humans and all other species agree, which is hard to imagine.

The most explicit critique of Bostrom's orthogonality thesis comes from the noted roboticist Rodney Brooks, who asserts that it's impossible for a program to be "smart enough that it would be able to invent ways to subvert human society to achieve goals set for it by humans, without understanding the ways in which it was causing problems for those same humans."[31] Unfortunately, it's not only possible for a program to behave like this; it is, in fact, inevitable, given the way Brooks defines the issue. Brooks posits that the optimal plan to "achieve goals set for it by humans" is causing problems for humans. It follows that those problems reflect things of value to humans that were omitted from the goals set for it by humans. The optimal plan being carried out by the machine may well cause problems for humans, and the machine may well be aware of this. But, by definition, the machine will not recognize those problems as problematic. They are none of its concern.

Steven Pinker seems to agree with Bostrom's orthogonality thesis, writing that "intelligence is the ability to deploy novel means to attain a goal; the goals are extraneous to the intelligence itself."[32] On the other hand, he finds it inconceivable that "the AI would be so brilliant that it could figure out how to transmute elements and rewire brains, yet so imbecilic that it would wreak havoc based on elementary blunders of misunderstanding."[33] He continues, "The ability to choose an action that best satisfies conflicting goals is not an add-on that engineers might forget to install and test; it *is* intelligence. So is the ability to interpret the intentions of a language user in context." Of course, "satisf[ying] conflicting goals" is not the problem—that's something that's been built into the standard model from the early days of decision theory. The problem is that the conflicting goals of which the machine is aware do not constitute the entirety of human concerns;

moreover, within the standard model, there's nothing to say that the machine has to care about goals it's not told to care about.

There are, however, some useful clues in what Brooks and Pinker say. It does seem stupid *to us* for the machine to, say, change the color of the sky as a side effect of pursuing some other goal, while ignoring the obvious signs of human displeasure that result. It seems stupid to us because we are attuned to noticing human displeasure and (usually) we are motivated to avoid causing it—even if we were previously unaware that the humans in question cared about the color of the sky. That is, we humans (1) care about the preferences of other humans and (2) know that we don't know what all those preferences are. In the next chapter, I argue that these characteristics, when built into a machine, may provide the beginnings of a solution to the King Midas problem.

The Debate, Restarted

This chapter has provided a glimpse into an ongoing debate in the broad intellectual community, a debate between those pointing to the risks of AI and those who are skeptical about the risks. It has been conducted in books, blogs, academic papers, panel discussions, interviews, tweets, and newspaper articles. Despite their valiant efforts, the "skeptics"—those who argue that the risk from AI is negligible—have failed to explain why superintelligent AI systems will necessarily remain under human control; and they have not even tried to explain why superintelligent AI systems will never be developed.

Many skeptics will admit, if pressed, that there *is* a real problem, even if it's not imminent. Scott Alexander, in his *Slate Star Codex* blog, summed it up brilliantly:[34]

> The "skeptic" position seems to be that, although we should prob-
> ably get a couple of bright people to start working on preliminary

aspects of the problem, we shouldn't panic or start trying to ban AI research.

The "believers," meanwhile, insist that although we shouldn't panic or start trying to ban AI research, we should probably get a couple of bright people to start working on preliminary aspects of the problem.

Although I would be happy if the skeptics came up with an irrefutable objection, perhaps in the form of a simple and foolproof (and evil-proof) solution to the control problem for AI, I think it's quite likely that this isn't going to happen, any more than we're going to find a simple and foolproof solution for cybersecurity or a simple and foolproof way to generate nuclear energy with zero risk. Rather than continue the descent into tribal name-calling and repeated exhumation of discredited arguments, it seems better, as Alexander puts it, to start working on some preliminary aspects of the problem.

The debate has highlighted the conundrum we face: if we build machines to optimize objectives, the objectives we put into the machines have to match what we want, but we don't know how to define human objectives completely and correctly. Fortunately, there is a middle way.

AI: A DIFFERENT APPROACH

Once the skeptic's arguments have been refuted and all the *but but buts* have been answered, the next question is usually, "OK, I admit there's a problem, but there's no solution, is there?" Yes, there is a solution.

Let's remind ourselves of the task at hand: to design machines with a high degree of intelligence—so that they can help us with difficult problems—while ensuring that those machines never behave in ways that make us seriously unhappy.

The task is, fortunately, not the following: given a machine that possesses a high degree of intelligence, work out how to control it. If that were the task, we would be toast. A machine viewed as a black box, a *fait accompli*, might as well have arrived from outer space. And our chances of controlling a superintelligent entity from outer space are roughly zero. Similar arguments apply to methods of creating AI systems that guarantee we won't understand how they work; these methods include *whole-brain emulation*[1]—creating souped-up electronic copies of human brains—as well as methods based on simulated evolution of programs.[2] I won't say more about these proposals because they are so obviously a bad idea.

So, how has the field of AI approached the "design machines with a high degree of intelligence" part of the task in the past? Like many other fields, AI has adopted the standard model: we build optimizing machines, we feed objectives into them, and off they go. That worked well when the machines were stupid and had a limited scope of action; if you put in the wrong objective, you had a good chance of being able to switch off the machine, fix the problem, and try again.

As machines designed according to the standard model become more intelligent, however, and as their scope of action becomes more global, the approach becomes untenable. Such machines will pursue their objective, no matter how wrong it is; they will resist attempts to switch them off; and they will acquire any and all resources that contribute to achieving the objective. Indeed, the optimal behavior for the machine might include deceiving the humans into thinking they gave the machine a reasonable objective, in order to gain enough time to achieve the actual objective given to it. This wouldn't be "deviant" or "malicious" behavior requiring consciousness and free will; it would just be part of an optimal plan to achieve the objective.

In Chapter 1, I introduced the idea of beneficial machines—that is, machines whose actions can be expected to achieve *our* objectives rather than *their* objectives. My goal in this chapter is to explain in simple terms how this can be done, despite the apparent drawback that the machines don't know what our objectives are. The resulting approach should lead eventually to machines that present no threat to us, no matter how intelligent they are.

Principles for Beneficial Machines

I find it helpful to summarize the approach in the form of three[3] principles. When reading these principles, keep in mind that they are intended primarily as a guide to AI researchers and developers in thinking about how to create beneficial AI systems; they are *not* intended as explicit laws for AI systems to follow:[4]

1. The machine's only objective is to maximize the realization of human preferences.

2. The machine is initially uncertain about what those preferences are.

3. The ultimate source of information about human preferences is human behavior.

Before delving into more detailed explanations, it's important to remember the broad scope of what I mean by *preferences* in these principles. Here's a reminder of what I wrote in Chapter 2: *if you were somehow able to watch two movies, each describing in sufficient detail and breadth a future life you might lead, such that each constitutes a virtual experience, you could say which you prefer, or express indifference.* Thus, preferences here are all-encompassing; they cover everything you might care about, arbitrarily far into the future.[5] And they are yours: the machine is not looking to identify or adopt one ideal set of preferences but to understand and satisfy (to the extent possible) the preferences of each person.

The first principle: Purely altruistic machines

The first principle, that the machine's only objective is to maximize the realization of human preferences, is central to the notion of a beneficial machine. In particular, it will be beneficial *to humans*, rather than to, say, cockroaches. There's no getting around this recipient-specific notion of benefit.

The principle means that the machine is purely altruistic—that is, it attaches absolutely no intrinsic value to its own well-being or even its own existence. It might protect itself in order to continue doing useful things for humans, or because its owner would be unhappy about having to pay for repairs, or because the sight of a dirty or damaged robot might be mildly distressing to passersby, but not because it wants to be alive. Putting in any preference for self-preservation sets up an additional incentive within the robot that is not strictly aligned with human well-being.

The wording of the first principle brings up two questions of fundamental importance. Each merits an entire bookshelf to itself, and in fact many books have already been written on these questions.

The first question is whether humans really have preferences in a meaningful or stable sense. In truth, the notion of a "preference" is an idealization that fails to match reality in several ways. For example, we aren't born with the preferences we have as adults, so they must change over time. For now, I will assume that the idealization is reasonable. Later, I will examine what happens when we give up the idealization.

The second question is a staple of the social sciences: given that it is usually impossible to ensure that everyone gets their most preferred outcome—we can't all be Emperor of the Universe—how should the machine trade off the preferences of multiple humans? Again, for the time being—and I promise to return to this question in the next chapter—it seems reasonable to adopt the simple approach of treating everyone equally. This is reminiscent of the roots of eighteenth-century utilitarianism in the phrase "the greatest happiness for the greatest numbers,"[6] and there are many caveats and elaborations required to make this work in practice. Perhaps the most important of these is the matter of the possibly vast number of people not yet born, and how their preferences are to be taken into account.

The issue of future humans brings up another, related question: How do we take into account the preferences of nonhuman entities? That is, should the first principle include the preferences of animals? (And possibly plants too?) This is a question worthy of debate, but the outcome seems unlikely to have a strong impact on the path forward for AI. For what it's worth, human preferences can and do include terms for the well-being of animals, as well as for the aspects of human well-being that benefit directly from animals' existence.[7] To say that the machine should pay attention to the preferences of animals *in addition* to this is to say that humans should build machines that care more about animals than humans do, which is a difficult position to sustain. A more tenable position is that our tendency to engage in myopic decision making—which works against our own interests—often leads to negative consequences for the environment and its animal inhabitants. A machine that makes less myopic decisions would help humans adopt more environmentally sound policies. And if, in the future, we give substantially greater weight to the well-being of animals than we currently do—which probably means sacrificing some of our own intrinsic well-being—then machines will adapt accordingly.

The second principle: Humble machines

The second principle, that the machine is initially *uncertain* about what human preferences are, is the key to creating beneficial machines.

A machine that assumes it knows the true objective perfectly will pursue it single-mindedly. It will never ask whether some course of action is OK, because it already knows it's an optimal solution for the objective. It will ignore humans jumping up and down screaming, "Stop, you're going to destroy the world!" because those are just words. Assuming perfect knowledge of the objective decouples the machine from the human: what the human does no longer matters, because the machine knows the goal and pursues it.

On the other hand, a machine that is uncertain about the true objective will exhibit a kind of humility: it will, for example, defer to humans and allow itself to be switched off. It reasons that the human will switch it off only if it's doing something wrong—that is, doing something contrary to human preferences. By the first principle, it wants to avoid doing that, but, by the second principle, it knows that's possible because it doesn't know exactly what "wrong" is. So, if the human does switch the machine off, then the machine avoids doing the wrong thing, and that's what it wants. In other words, the machine has a positive incentive to allow itself to be switched off. It remains coupled to the human, who is a potential source of information that will allow it to avoid mistakes and do a better job.

Uncertainty has been a central concern in AI since the 1980s; indeed the phrase "modern AI" often refers to the revolution that took place when uncertainty was finally recognized as a ubiquitous issue in real-world decision making. Yet uncertainty in the *objective* of the AI system was simply ignored. In all the work on utility maximization, goal achievement, cost minimization, reward maximization, and loss minimization, it is assumed that the utility function, the goal, the cost function, the reward function, and the loss function are known perfectly. How could this be? How could the AI community (and the control theory, operations research, and statistics communities) have such a huge blind spot for so long, even while embracing uncertainty in all other aspects of decision making?[8]

One could make some rather complicated technical excuses,[9] but I suspect the truth is that, with some honorable exceptions,[10] AI researchers simply bought into the standard model that maps our notion of human intelligence onto machine intelligence: humans have objectives and pursue them, so machines should have objectives and pursue them. They, or should I say we, never really examined this fundamental assumption. It is built into all existing approaches for constructing intelligent systems.

The third principle: Learning to predict human preferences

The third principle, that the ultimate source of information about human preferences is human behavior, serves two purposes.

The first purpose is to provide a definite grounding for the term *human preferences*. By assumption, human preferences aren't in the machine and it cannot observe them directly, but there must still be some definite connection between the machine and human preferences. The principle says that the connection is through the observation of human *choices*: we assume that choices are related in some (possibly very complicated) way to underlying preferences. To see why this connection is essential, consider the converse: if some human preference had *no effect whatsoever* on any actual or hypothetical choice the human might make, then it would probably be meaningless to say that the preference exists.

The second purpose is to enable the machine to become more useful as it learns more about what we want. (After all, if it knew *nothing* about human preferences, it would be of no use to us.) The idea is simple enough: human choices reveal information about human preferences. Applied to the choice between pineapple pizza and sausage pizza, this is straightforward. Applied to choices between future lives and choices made with the goal of influencing the robot's behavior, things get more interesting. In the next chapter I explain how to formulate and solve such problems. The real complications arise, however, because humans are not perfectly rational: imperfection comes between human preferences and human choices, and the machine must take into account those imperfections if it is to interpret human choices as evidence of human preferences.

Not what I mean

Before going into more detail, I want to head off some potential misunderstandings.

The first and most common misunderstanding is that I am proposing to install in machines a single, idealized value system of my own design that guides the machine's behavior. "Whose values are you going to put in?" "Who gets to decide what the values are?" Or even, "What gives Western, well-off, white male cisgender scientists such as Russell the right to determine how the machine encodes and develops human values?"[11]

I think this confusion comes partly from an unfortunate conflict between the commonsense meaning of *value* and the more technical sense in which it is used in economics, AI, and operations research. In ordinary usage, values are what one uses to help resolve moral dilemmas; as a technical term, on the other hand, *value* is roughly synonymous with utility, which measures the degree of desirability of anything from pizza to paradise. The meaning I want is the technical one: I just want to make sure the machines give me the right pizza and don't accidentally destroy the human race. (Finding my keys would be an unexpected bonus.) To avoid this confusion, the principles talk about human *preferences* rather than human *values*, since the former term seems to steer clear of judgmental preconceptions about morality.

"Putting in values" is, of course, exactly the mistake I am saying we should avoid, because getting the values (or preferences) exactly right is so difficult and getting them wrong is potentially catastrophic. I am proposing instead that machines learn to predict better, for each person, which life that person would prefer, all the while being aware that the predictions are highly uncertain and incomplete. In principle, the machine can learn billions of different predictive preference models, one for each of the billions of people on Earth. This is really not too much to ask for the AI systems of the future, given that present-day Facebook systems are already maintaining more than two billion individual profiles.

A related misunderstanding is that the goal is to equip machines with "ethics" or "moral values" that will enable them to resolve moral dilemmas. Often, people bring up the so-called trolley problems,[12] where one has to choose whether to kill one person in order to save others, because of their supposed relevance to self-driving cars. The whole point of moral dilemmas, however, is that they are dilemmas: there are good arguments on both sides. The survival of the human race is not a moral dilemma. Machines could solve most moral dilemmas the *wrong* way (whatever that is) and still have no catastrophic impact on humanity.[13]

Another common supposition is that machines that follow the three principles will adopt all the sins of the evil humans they observe and learn from. Certainly, there are many of us whose choices leave something to be desired, but there is no reason to suppose that machines who study our motivations will make the same choices, any more than criminologists become criminals. Take, for example, the corrupt government official who demands bribes to approve building permits because his paltry salary won't pay for his children to go to university. A machine observing this behavior will not learn to take bribes; it will learn that the official, like many other people, has a very strong desire for his children to be educated and successful. It will find ways to help him that don't involve lowering the well-being of others. This is not to say that *all* cases of evil behavior are unproblematic for machines—for example, machines may need to treat differently those who actively prefer the suffering of others.

Reasons for Optimism

In a nutshell, I am suggesting that we need to steer AI in a radically new direction if we want to retain control over increasingly intelligent machines. We need to move away from one of the driving ideas of twentieth-century technology: machines that optimize a given objective. I am often asked why I think this is even remotely feasible, given the huge momentum behind the standard model in AI and related disciplines. In fact, I am quite optimistic that it can be done.

The first reason for optimism is that there are strong economic incentives to develop AI systems that defer to humans and gradually align themselves to user preferences and intentions. Such systems will be highly desirable: the range of behaviors they can exhibit is simply far greater than that of machines with fixed, known objectives. They will ask humans questions or ask for permission when appropriate; they will do "trial runs" to see if we like what they propose to do; they will accept correction when they do something wrong. On the other hand, systems that fail to do this will have severe consequences. Up to now, the stupidity and limited scope of AI systems has protected us from these consequences, but that will change. Imagine, for example, some future domestic robot charged with looking after your children while you are working late. The children are hungry, but the refrigerator is empty. Then the robot notices the cat. Alas, the robot understands the cat's nutritional value but not its sentimental value. Within a few short hours, headlines about deranged robots and roasted cats are blanketing the world's media and the entire domestic-robot industry is out of business.

The possibility that one industry player could destroy the entire industry through careless design provides a strong economic motivation to form safety-oriented industry consortia and to enforce safety standards. Already, the Partnership on AI, which includes as members nearly all the world's leading technology companies, has agreed to cooperate to ensure that "AI research and technology is robust, reliable, trustworthy, and operates within secure constraints." To my knowledge, all the major players are publishing their safety-oriented research in the open literature. Thus, the economic incentive is in operation long before we reach human-level AI and will only strengthen over time. Moreover, the same cooperative dynamic may be starting at the international level—for example, the stated policy of the Chinese government is to "cooperate to preemptively prevent the threat of AI."[14]

A second reason for optimism is that the raw data for learning about human preferences—namely, examples of human behavior—are so abundant. The data come not just in the form of direct observation via camera, keyboard, and touch screen by billions of machines sharing data with one another about billions of humans (subject to privacy constraints, of course) but also in indirect form. The most obvious kind of indirect evidence is the vast human record of books, films, and television and radio broadcasts, which is almost entirely concerned with *people doing things* (and other people being upset about it). Even the earliest and most tedious Sumerian and Egyptian records of copper ingots being traded for sacks of barley give some insight into human preferences for different commodities.

There are, of course, difficulties involved in interpreting this raw material, which includes propaganda, fiction, the ravings of lunatics, and even the pronouncements of politicians and presidents, but there is certainly no reason for the machine to take it all at face value. Machines can and should interpret all communications from other intelligent entities as moves in a game rather than as statements of fact; in some games, such as cooperative games with one human and one machine, the human has an incentive to be truthful, but in many other situations there are incentives to be dishonest. And of course, whether honest or dishonest, humans may be deluded in their own beliefs.

There is a second kind of indirect evidence that is staring us in the face: the way we have made the world.[15] We made it that way because—very roughly—we like it that way. (Obviously, it's not perfect!) Now, imagine you are an alien visiting Earth while all the humans are away on holiday. As you peer inside their houses, can you begin to grasp the basics of human preferences? Carpets are on floors because we like to walk on soft, warm surfaces and we don't like loud footsteps; vases are on the middle of the table rather than the edge because we don't want them to fall and break; and so on—everything that isn't arranged by nature itself provides clues to the likes and dislikes of the strange bipedal creatures who inhabit this planet.

Reasons for Caution

You may find the Partnership on AI's promises of cooperation on AI safety less than reassuring if you have been following progress in self-driving cars. That field is ruthlessly competitive, for some very good reasons: the first car manufacturer to release a fully autonomous vehicle will gain a huge market advantage; that advantage will be self-reinforcing because the manufacturer will be able to collect more data more quickly to improve the system's performance; and ride-hailing companies such as Uber would quickly go out of business if another company were to roll out fully autonomous taxis before Uber does. This has led to a high-stakes race in which caution and careful engineering appear to be less important than snazzy demos, talent grabs, and premature rollouts.

Thus, life-or-death economic competition provides an impetus to cut corners on safety in the hope of winning the race. In a 2008 retrospective paper on the 1975 Asilomar conference that he co-organized—the conference that led to a moratorium on genetic modification of humans—the biologist Paul Berg wrote,[16]

> There is a lesson in Asilomar for all of science: the best way to respond to concerns created by emerging knowledge or early-stage technologies is for scientists from publicly funded institutions to find common cause with the wider public about the best way to regulate—as early as possible. Once scientists from corporations begin to dominate the research enterprise, it will simply be too late.

Economic competition occurs not just between corporations but also between nations. A recent flurry of announcements of multibillion-dollar national investments in AI from the United States, China, France, Britain, and the EU certainly suggests that none of the major powers wants to be left behind. In 2017, Russian president Vladimir Putin said, "The one who becomes the leader in [AI] will be the ruler of the world."[17] This analysis is essentially correct. Advanced AI would, as we saw in Chapter 3, lead to greatly increased productivity and rates of innovation in almost all areas. If not shared, it would allow its possessor to outcompete any rival nation or bloc.

Nick Bostrom, in *Superintelligence*, warns against exactly this motivation. National competition, just like corporate competition, would tend to focus more on advances in raw capabilities and less on the problem of control. Perhaps, however, Putin has read Bostrom; he went on to say, "It would be strongly undesirable if someone wins a monopolist position." It would also be rather pointless, because human-level AI is not a zero-sum game and nothing is lost by sharing it. On the other hand, competing to be the first to achieve human-level AI, without first solving the control problem, is a negative-sum game. The payoff for everyone is minus infinity.

There's only a limited amount that AI researchers can do to influence the evolution of global policy on AI. We can point to possible applications that would provide economic and social benefits; we can warn about possible misuses such as surveillance and weapons; and we can provide roadmaps for the likely path of future developments and their impacts. Perhaps the most important thing we can do is to design AI systems that are, to the extent possible, provably safe and beneficial for humans. Only then will it make sense to attempt general regulation of AI.

$$\boxed{8}$$

PROVABLY BENEFICIAL AI

I f we are going to rebuild AI along new lines, the foundations must be solid. When the future of humanity is at stake, hope and good intentions—and educational initiatives and industry codes of conduct and legislation and economic incentives to do the right thing— are not enough. All of these are fallible, and they often fail. In such situations, we look to precise definitions and rigorous step-by-step mathematical proofs to provide incontrovertible guarantees.

That's a good start, but we need more. We need to be sure, to the extent possible, that what is guaranteed is actually what we want and that the assumptions going into the proof are actually true. The proofs themselves belong in journal papers written for specialists, but I think it is useful nonetheless to understand what proofs are and what they can and cannot provide in the way of real safety. The "provably beneficial" in the title of the chapter is an aspiration rather than a promise, but it is the right aspiration.

Mathematical Guarantees

We will want, eventually, to prove theorems to the effect that a particular way of designing AI systems ensures that they will be beneficial to humans. A theorem is just a fancy name for an assertion, stated precisely enough so that its truth in any particular situation can be checked. Perhaps the most famous theorem is Fermat's Last Theorem, which was conjectured by the French mathematician Pierre de Fermat in 1637 and finally proved by Andrew Wiles in 1994 after 357 years of effort (not all of it by Wiles).[1] The theorem can be written in one line, but the proof is over one hundred pages of dense mathematics.

Proofs begin from *axioms*, which are assertions whose truth is simply assumed. Often, the axioms are just definitions, such as the definitions of integers, addition, and exponentiation needed for Fermat's theorem. The proof proceeds from the axioms by logically incontrovertible steps, adding new assertions until the theorem itself is established as a consequence of one of the steps.

Here's a fairly obvious theorem that follows almost immediately from the definitions of integers and addition: $1 + 2 = 2 + 1$. Let's call this *Russell's theorem*. It's not much of a discovery. On the other hand, Fermat's Last Theorem feels like something completely new—a discovery of something previously unknown. The difference, however, is just a matter of degree. The truth of both Russell's and Fermat's theorems is *already contained in the axioms*. Proofs merely make explicit what was already implicit. They can be long or short, but they add nothing new. The theorem is only as good as the assumptions that go into it.

That's fine when it comes to mathematics, because mathematics is about abstract objects that *we* define—numbers, sets, and so on. The axioms are true because we say so. On the other hand, if you want to prove something about the real world—for example, that AI systems designed like *so* won't kill you on purpose—your axioms have to be true in the real world. If they aren't true, you've proved something about an imaginary world.

Science and engineering have a long and honorable tradition of proving results about imaginary worlds. In structural engineering, for example, one might see a mathematical analysis that begins, "Let AB be a rigid beam. . . ." The word *rigid* here doesn't mean "made of something hard like steel"; it means "infinitely strong," so that it doesn't bend at all. Rigid beams do not exist, so this is an imaginary world. The trick is to know how far one can stray from the real world and still obtain useful results. For example, if the rigid-beam assumption allows an engineer to calculate the forces in a structure that includes the beam, and those forces are small enough to bend a real steel beam by only a tiny amount, then the engineer can be reasonably confident that the analysis will transfer from the imaginary world to the real world.

A good engineer develops a sense for when this transfer might fail—for example, if the beam is under compression, with huge forces pushing on it from each end, then even a tiny amount of bending might lead to greater lateral forces causing more bending, and so on, resulting in catastrophic failure. In that case, the analysis is redone with "Let AB be a flexible beam with stiffness K. . . ." This is still an imaginary world, of course, because real beams do not have uniform stiffness; instead, they have microscopic imperfections that can lead to cracks forming if the beam is subject to repeated bending. The process of removing unrealistic assumptions continues until the engineer is fairly confident that the remaining assumptions are true enough in the real world. After that, the engineered system can be tested in the real world; but the test results are just that. They do not prove that the same system will work in other circumstances or that other instances of the system will behave the same way as the original.

One of the classic examples of assumption failure in computer science comes from cybersecurity. In that field, a huge amount of mathematical analysis goes into showing that certain digital protocols are *provably secure*—for example, when you type a password into a Web application, you want to be sure that it is encrypted before transmission so that someone eavesdropping on the network cannot read your password. Such digital systems are often provably secure but still vulnerable to attack in reality. The false assumption here is that this is a digital process. It isn't. It operates in the real, physical world. By listening to the sound of your keyboard or measuring voltages on the electrical line that supplies power to your desktop computer, an attacker can "hear" your password or observe the encryption/decryption calculations that are occurring as it is processed. The cybersecurity community is now responding to these so-called side-channel attacks— for example, by writing encryption code that produces the same voltage fluctuations regardless of what message is being encrypted.

Let's look at the kind of theorem we would like eventually to prove about machines that are beneficial to humans. One type might go something like this:

> Suppose a machine has components A, B, C, connected to each other like *so* and to the environment like *so*, with internal learning algorithms l_A, l_B, l_C that optimize internal feedback rewards r_A, r_B, r_C defined like *so*, and [a few more conditions] . . . then, with very high probability, the machine's behavior will be very close in value (for humans) to the best possible behavior realizable on any machine with the same computational and physical capabilities.

The main point here is that such a theorem should hold *regardless of how smart the components become*—that is, the vessel never springs a leak and the machine always remains beneficial to humans.

There are three other points worth making about this kind of theorem. First, we cannot try to prove that the machine produces optimal (or even near-optimal) behavior on our behalf, because that's almost certainly computationally impossible. For example, we might want the machine to play Go perfectly, but there is good reason to believe that cannot be done in any practical amount of time on any physically realizable machine. Optimal behavior in the real world is even less feasible. Hence, the theorem says "best possible" rather than "optimal."

Second, we say "very high probability . . . very close" because that's typically the best that can be done with machines that learn. For example, if the machine is learning to play roulette for us and the ball lands in zero forty times in a row, the machine might reasonably decide the table was rigged and bet accordingly. But it *could* have happened by chance; so there is always a small—perhaps vanishingly small—chance of being misled by freak occurrences. Finally, we are a long way from being able to prove any such theorem for really intelligent machines operating in the real world!

There are also analogs of the side-channel attack in AI. For example, the theorem begins with "Suppose a machine has components *A*, *B*, *C*, connected to each other like *so.* . . ." This is typical of all correctness theorems in computer science: they begin with a description of the program being proved correct. In AI, we typically distinguish between the *agent* (the program doing the deciding) and the *environment* (on which the agent acts). Since we design the agent, it seems reasonable to assume that it has the structure we give it. To be extra safe, we can prove that its learning processes can modify its program only in certain circumscribed ways that cannot cause problems. Is this enough? No. As with side-channel attacks, the assumption that the program operates within a digital system is incorrect. Even if a learning algorithm is constitutionally incapable of overwriting its own code by digital means, it may, nonetheless, learn to persuade humans to do "brain surgery" on it—to violate the agent/environment distinction and change the code by physical means.[2]

Unlike the structural engineer reasoning about rigid beams, we have very little experience with the assumptions that will eventually underlie theorems about provably beneficial AI. In this chapter, for example, we will typically be assuming a rational human. This is a bit like assuming a rigid beam, because there are no perfectly rational humans in reality. (It's probably much worse, however, because humans are not even close to being rational.) The theorems we can prove seem to provide some insights, and the insights survive the introduction of a certain degree of randomness in human behavior, but it is as yet far from clear what happens when we consider some of the complexities of real humans.

So, we are going to have to be very careful in examining our assumptions. When a proof of safety succeeds, we need to make sure it's not succeeding because we have made unrealistically strong assumptions or because the definition of safety is too weak. When a proof of safety fails, we need to resist the temptation to strengthen the assumptions to make the proof go through—for example, by adding the assumption that the program's code remains fixed. Instead, we need to tighten up the design of the AI system—for example, by ensuring that it has no incentive to modify critical parts of its own code.

There are some assumptions that I call OWMAWGH assumptions, standing for "otherwise we might as well go home." That is, if these assumptions are false, the game is up and there is nothing to be done. For example, it is reasonable to assume that the universe operates according to constant and somewhat discernible laws. If this is not the case, we will have no assurance that learning processes—even very sophisticated ones—will work at all. Another basic assumption is that humans care about what happens; if not, provably beneficial AI has no purpose because *beneficial* has no meaning. Here, *caring* means having roughly coherent and more-or-less stable preferences about the future. In the next chapter, I examine the consequences of *plasticity* in human preferences, which presents a serious philosophical challenge to the very idea of provably beneficial AI.

For now, I focus on the simplest case: a world with one human and one robot. This case serves to introduce the basic ideas, but it's also useful in its own right: you can think of the human as standing in for all of humanity and the robot as standing in for all machines. Additional complications arise when considering multiple humans and machines.

Learning Preferences from Behavior

Economists elicit preferences from human subjects by offering them choices.[3] This technique is widely used in product design, marketing, and interactive e-commerce systems. For example, by offering test subjects choices among cars with different paint colors, seating arrangements, trunk sizes, battery capacities, cup holders, and so on, a car designer learns how much people care about various car features and how much they are willing to pay for them. Another important application is in the medical domain, where an oncologist considering a possible limb amputation might want to assess the patient's preferences between mobility and life expectancy. And of course, pizza restaurants want to know how much more someone is willing to pay for sausage pizza than plain pizza.

Preference elicitation typically considers only single choices made between objects whose value is assumed to be immediately apparent to the subject. It's not obvious how to extend it to preferences between future lives. For that, we (and machines) need to learn from observations of behavior over time—behavior that involves multiple choices and uncertain outcomes.

Early in 1997, I was involved in discussions with my colleagues Michael Dickinson and Bob Full about ways in which we might be able to apply ideas from machine learning to understand the locomotive behavior of animals. Michael studied in exquisite detail the wing motions of fruit flies. Bob was especially fond of creepy-crawlies and had built a little treadmill for cockroaches to see how their gait changed with speed. We thought it might be possible to use reinforcement learning to train a robotic or simulated insect to reproduce these complex behaviors. The problem we faced was that we didn't know what reward signal to use. What were the flies and cockroaches optimizing? Without that information, we couldn't apply reinforcement learning to train the virtual insect, so we were stuck.

One day, I was walking down the road that leads from our house in Berkeley to the local supermarket. The road has a downhill slope, and I noticed, as I am sure most people have, that the slope induced a slight change in the way I walked. Moreover, the uneven paving resulting from decades of minor earthquakes induced additional gait changes, including raising my feet a little higher and planting them less stiffly because of the unpredictable ground level. As I pondered these mundane observations, I realized we had got it backwards. While reinforcement learning generates behavior from rewards, we actually wanted the opposite: to learn the rewards given the behavior. We already had the behavior, as produced by the flies and cockroaches; we wanted to know the specific reward signal being optimized by this behavior. In other words, we needed algorithms for *inverse* reinforcement learning, or IRL.[4] (I did not know at the time that a similar problem had been studied under the perhaps less wieldy name of *structural estimation of Markov decision processes*, a field pioneered by Nobel laureate Tom Sargent in the late 1970s.[5]) Such algorithms would not only be able to explain animal behavior but also to predict their behavior in new circumstances. For example, how would a cockroach run on a bumpy treadmill that sloped sideways?

The prospect of answering such fundamental questions was almost too exciting to bear, but even so it took some time to work out the first algorithms for IRL.[6] Many different formulations and algorithms for IRL have been proposed since then. There are formal guarantees that the algorithms work, in the sense that they can acquire enough information about an entity's preferences to be able to behave just as successfully as the entity they are observing.[7]

Perhaps the easiest way to understand IRL is this: the observer starts with some vague estimate of the true reward function and then refines this estimate, making it more precise, as more behavior is observed. Or, in Bayesian language:[8] start with a prior probability over possible reward functions and then update the probability distribution on reward functions as evidence arrives.[c] For example, suppose Robbie the robot is watching Harriet the human and wondering how much she prefers aisle seats to window seats. Initially, he is quite uncertain about this. Conceptually, Robbie's reasoning might go like this: "If Harriet really cared about an aisle seat, she would have looked at the seat map to see if one was available rather than just accepting the window seat that the airline gave her, but she didn't, even though she probably noticed it was a window seat and she probably wasn't in a hurry; so now it's considerably more likely that she either is roughly indifferent between window and aisle or even prefers a window seat."

The most striking example of IRL in practice is the work of my colleague Pieter Abbeel on learning to do helicopter aerobatics.[9] Expert human pilots can make model helicopters do amazing things—loops, spirals, pendulum swings, and so on. Trying to copy what the human *does* turns out not to work very well because conditions are not perfectly reproducible: repeating the same control sequences in different circumstances can lead to disaster. Instead, the algorithm learns what the human pilot *wants*, in the form of trajectory constraints that it can achieve. This approach actually produces results that are even better than the human expert's, because the human has slower reactions and is constantly making small mistakes and correcting for them.

Assistance Games

IRL is already an important tool for building effective AI systems, but it makes some simplifying assumptions. The first is that the robot is going to *adopt* the reward function once it has learned it by observing the human, so that it can perform the same task. This is fine for driving or helicopter piloting, but it's not fine for drinking coffee: a robot observing my morning routine should learn that I (sometimes) want coffee, but should not learn to want coffee itself. Fixing this issue is easy—we simply ensure that the robot associates the preferences with the human, not with itself.

The second simplifying assumption in IRL is that the robot is observing a human who is solving a single-agent decision problem. For example, suppose the robot is in medical school, learning to be a surgeon by watching a human expert. IRL algorithms assume that the human performs the surgery in the usual optimal way, as if the robot were not there. But that's not what would happen: the human surgeon is motivated to have the robot (like any other medical student) learn quickly and well, and so she will modify her behavior considerably. She might explain what she is doing as she goes along; she might point out mistakes to avoid, such as making the incision too deep or the stitches too tight; she might describe the contingency plans in case something goes wrong during surgery. None of these behaviors make sense when performing surgery in isolation, so IRL algorithms will not be able to interpret the preferences they imply. For this reason, we will need to generalize IRL from the single-agent setting to the multi-agent setting—that is, we will need to devise learning algorithms that work when the human and robot are part of the same environment and interacting with each other.

With a human and a robot in the same environment, we are in the realm of game theory—just as in the penalty shoot-out between Alice and Bob on page 28. We assume, in this first version of the theory, that the human has preferences and acts according to those preferences. The robot doesn't know what preferences the human has, but it wants to satisfy them anyway. We'll call any such situation an *assistance game*, because the robot is, by definition, supposed to be helpful to the human.[10]

Assistance games instantiate the three principles from the preceding chapter: the robot's only objective is to satisfy human preferences, it doesn't initially know what they are, and it can learn more by observing human behavior. Perhaps the most interesting property of assistance games is that, by solving the game, the robot can work out for itself how to interpret the human's behavior as providing information about human preferences.

The paperclip game

The first example of an assistance game is the paperclip game. It's a very simple game in which Harriet the human has an incentive to "signal" to Robbie the robot some information about her preferences. Robbie is able to interpret that signal because he can solve the game, and therefore he can understand what would have to be true about Harriet's preferences in order for her to signal in that way.

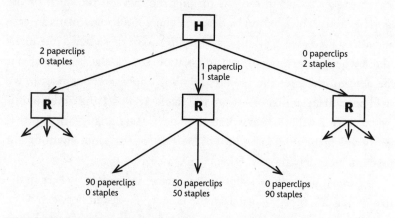

FIGURE 12: The paperclip game. Harriet the human can choose to make 2 paperclips, 2 staples, or 1 of each. Robbie the robot then has a choice to make 90 paperclips, 90 staples, or 50 of each.

The steps of the game are depicted in figure 12. It involves making paperclips and staples. Harriet's preferences are expressed by a payoff function that depends on the number of paperclips and the number of staples produced, with a certain "exchange rate" between the two. For example, she might value paperclips at 45¢ and staples at 55¢ each. (We'll assume the two values always add up to $1.00; it's only the ratio that matters.) So, if 10 paperclips and 20 staples are produced, Harriet's payoff will be 10 × 45¢ + 20 × 55¢ = $15.50. Robbie the robot is initially completely uncertain about Harriet's preferences: he has a uniform distribution for the value of a paperclip (that is, it's equally likely to be any value from 0¢ to $1.00). Harriet goes first and can choose to make two paperclips, two staples, or one of each. Then Robbie can choose to make 90 paperclips, 90 staples, or 50 of each.[11]

Notice that if she were doing this by herself, Harriet would just make two staples, with a value of $1.10. But Robbie is watching, and he learns from her choice. What exactly does he learn? Well, that depends on how Harriet makes her choice. How does Harriet make her choice? That depends on how Robbie is going to interpret it. So, we seem to have a circular problem! That's typical in game-theoretic problems, and that's why Nash proposed the concept of equilibrium solutions.

To find an equilibrium solution, we need to identify strategies for Harriet and Robbie such that neither has an incentive to change their strategy, assuming the other remains fixed. A strategy for Harriet specifies how many paperclips and staples to make, given her preferences; a strategy for Robbie specifies how many paperclips and staples to make, given Harriet's action.

It turns out there is only one equilibrium solution, and it looks like this:

- Harriet decides as follows based on her value for paperclips:
  - *If the value is less than 44.6¢, make 0 paperclips and 2 staples.*
  - *If the value is between 44.6¢ and 55.4¢, make 1 of each.*
  - *If the value is more than 55.4¢, make 2 paperclips and 0 staples.*
- Robbie responds as follows:
  - *If Harriet makes 0 paperclips and 2 staples, make 90 staples.*

- *If Harriet makes 1 of each, make 50 of each.*
- *If Harriet makes 2 paperclips and 0 staples, make 90 paperclips.*

(In case you are wondering exactly how the solution is obtained, the details are in the notes.[12]) With this strategy, Harriet is, in effect, *teaching* Robbie about her preferences using a simple code—a language, if you like—that emerges from the equilibrium analysis. As in the example of surgical teaching, a single-agent IRL algorithm wouldn't understand this code. Note also that Robbie never learns Harriet's preferences exactly, but he learns enough to act optimally on her behalf—that is, he acts just as he would if he *did* know her preferences exactly. He is provably beneficial to Harriet under the assumptions stated and under the assumption that Harriet is playing the game correctly.

One can also construct problems where, like a good student, Robbie will ask questions, and, like a good teacher, Harriet will show Robbie the pitfalls to avoid. These behaviors occur not because we write scripts for Harriet and Robbie to follow, but because they are the optimal solution to the assistance game in which Harriet and Robbie are participants.

The off-switch game

An instrumental goal is one that is generally useful as a subgoal of almost any original goal. Self-preservation is one of these instrumental goals, because very few original goals are better achieved when dead. This leads to the *off-switch problem*: a machine that has a fixed objective will not allow itself to be switched off and has an incentive to disable its own off-switch.

The off-switch problem is really the core of the problem of control for intelligent systems. If we cannot switch a machine off because it won't let us, we're really in trouble. If we can, then we may be able to control it in other ways too.

It turns out that uncertainty about the objective is essential for ensuring that we can switch the machine off—even when it's more intelligent than us. We saw the informal argument in the previous chapter: by the first principle of beneficial machines, Robbie cares only about Harriet's preferences, but, by the second principle, he's unsure about what they are. He knows he doesn't want to do the wrong thing, but he doesn't know what that means. Harriet, on the other hand, does know (or so we assume, in this simple case). Therefore, if she switches Robbie off it's to avoid him doing something wrong, so he's happy to be switched off.

To make this argument more precise, we need a formal model of the problem.[13] I'll make it as simple as possible, but no simpler (see figure 13).

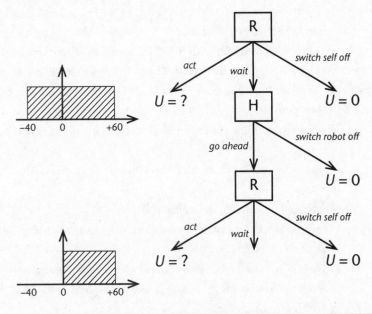

FIGURE 13: The off-switch game. Robbie can choose to act now, with a highly uncertain payoff; to commit suicide; or to wait for Harriet. Harriet can switch Robbie off or let him go ahead. Robbie now has the same choice again. Acting still has an uncertain payoff to Harriet, but now Robbie knows the payoff is not negative.

Robbie, now working as Harriet's personal assistant, has the first choice. He can act now—let's say he can book Harriet into an expensive hotel. He's quite unsure how much Harriet will like the hotel and its price—let's say he has a uniform probability for its net value to Harriet between −40 and +60, with an average of +10. He could also "switch himself off"—less melodramatically, take himself out of the hotel booking process altogether—which we define to have value 0 to Harriet. If those were his two choices, he would go ahead and book the hotel, incurring a significant risk of making Harriet unhappy. (If the range were −60 to +40, with an average of −10, he'd switch himself off.) We'll give Robbie a third choice, however: explain his plan, wait, and let Harriet switch him off. Harriet can either switch him off or let him go ahead and book the hotel. What possible good could this do, you may ask, given that he could make both of those choices himself?

The point is that Harriet's choice—to switch Robbie off or let him go ahead—provides Robbie with new information about Harriet's preferences. If Harriet lets Robbie go ahead, it's because the value to Harriet is positive. Now Robbie's belief is uniform between 0 and 60, with an average of 30.

So, if we evaluate Robbie's initial choices from his point of view:

- Acting now and booking the hotel has an expected value of +10.
- Switching himself off has a value of 0.
- Waiting and letting Harriet switch him off (if she so desires) leads to two possible outcomes:
  - *There is a 40 percent chance (based on Robbie's uncertainty about the hotel plan) that Harriet will hate it and will switch Robbie off, with value 0.*
  - *There's a 60 percent chance Harriet will like it and allow Robbie to go ahead, with expected value +30.*
- Thus, waiting has expected value 40% × 0 + 60% × 30 = +18, which is better than acting now at +10.

The upshot is that *Robbie has a positive incentive to allow himself to be switched off*. This incentive comes directly from Robbie's uncertainty about Harriet's preferences. Robbie is aware that there's a chance (40 percent in this example) that he might be about to do something that will make Harriet unhappy, in which case being switched off would be preferable to going ahead. Were Robbie already certain about Harriet's preferences, he would just go ahead and make the decision (or switch himself off). There would be absolutely nothing to be gained from consulting Harriet, because, according to Robbie's definite beliefs, he can already predict exactly what she is going to decide.

In fact, it is possible to prove the same result in the general case: as long as Robbie is not completely certain that he's about to do what Harriet herself would do, he will prefer to allow her to switch him off.[14] Her decision provides Robbie with information, and information is always useful for improving Robbie's decisions. Conversely, if Robbie is certain about Harriet's decision, her decision provides no new information, and so Robbie has no incentive to allow her to decide.

There are some obvious elaborations on the model that are worth exploring immediately. The first elaboration is to impose a cost for asking Harriet to make decisions or answer questions. (That is, we assume Robbie knows at least this much about Harriet's preferences: her time is valuable.) In that case, Robbie is less inclined to bother Harriet if he is nearly certain about her preferences; the larger the cost, the more uncertain Robbie has to be before bothering Harriet. This is as it should be. And if Harriet is *really* grumpy about being interrupted, she shouldn't be too surprised if Robbie occasionally does things she doesn't like.

The second elaboration is to allow for some probability of human error—that is, Harriet might sometimes switch Robbie off even when his proposed action is reasonable, and she might sometimes let Robbie go ahead even when his proposed action is undesirable. We can put this probability of human error into the mathematical model of the assistance game and find the solution, as before. As one might expect, the solution to the game shows that Robbie is less inclined to defer to an irrational Harriet who sometimes acts against her own best interests. The more randomly she behaves, the more uncertain Robbie has to be about her preferences before deferring to her. Again, this is as it should be—for example, if Robbie is an autonomous car and Harriet is his naughty two-year-old passenger, Robbie should *not* allow himself to be switched off by Harriet in the middle of the freeway.

There are many more ways in which the model can be elaborated or embedded into complex decision problems.[15] I am confident, however, that the core idea—the essential connection between helpful, deferential behavior and machine uncertainty about human preferences—will survive these elaborations and complications.

Learning preferences exactly in the long run

There is one important question that may have occurred to you in reading about the off-switch game. (Actually, you probably have *loads* of important questions, but I'm going to answer only this one.) What happens as Robbie acquires more and more information about Harriet's preferences, becoming less and less uncertain? Does that mean he will eventually stop deferring to her altogether? This is a ticklish question, and there are two possible answers: yes and yes.

The first *yes* is benign: as a general matter, as long as Robbie's initial beliefs about Harriet's preferences ascribe *some* probability, however small, to the preferences that she actually has, then as Robbie becomes more and more certain, he will become more and more right. That is, he will eventually be certain that Harriet has the preferences that she does in fact have. For example, if Harriet values paperclips at 12¢ and staples at 88¢, Robbie will eventually learn these values. In that case, Harriet doesn't care whether Robbie defers to her, because she knows he will always do exactly what she would have done in his place. There will never be an occasion where Harriet wants to switch Robbie off.

The second *yes* is less benign. If Robbie rules out, a priori, the true preferences that Harriet has, he will never learn those true preferences, but his beliefs may nonetheless converge to an incorrect assessment. In other words, over time, he becomes more and more certain about a false belief concerning Harriet's preferences. Typically, that false belief will be whichever hypothesis is closest to Harriet's true preferences, out of all the hypotheses that Robbie initially believes are possible. For example, if Robbie is absolutely certain that Harriet's value for paperclips lies between 25¢ and 75¢, and Harriet's true value is 12¢, then Robbie will eventually become certain that she values paperclips at 25¢.[16]

As he approaches certainty about Harriet's preferences, Robbie will resemble more and more the bad old AI systems with fixed objectives: he won't ask permission or give Harriet the option to turn him off, and he has the wrong objective. This is hardly dire if it's just paperclips versus staples, but it might be quality of life versus length of life if Harriet is seriously ill, or population size versus resource consumption if Robbie is supposedly acting on behalf of the human race.

We have a problem, then, if Robbie rules out in advance preferences that Harriet might in fact have: he may converge to a definite but incorrect belief about her preferences. The solution to this problem seems obvious: don't do it! Always allocate some probability, however small, to preferences that are logically possible. For example, it's logically possible that Harriet actively wants to get rid of staples and would pay you to take them away. (Perhaps as a child she stapled her finger to the table, and now she cannot stand the sight of them.) So, we should allow for negative exchange rates, which makes things a bit more complicated but still perfectly manageable.[17]

But what if Harriet values paperclips at 12¢ on weekdays and 80¢ on weekends? This new preference is not describable by any single number, and so Robbie has, in effect, ruled it out in advance. It's just not in his set of possible hypotheses about Harriet's preferences. More generally, there might be many, many things besides paperclips and staples that Harriet cares about. (Really!) Suppose, for example, that

Harriet is concerned about the climate, and suppose that Robbie's initial belief allows for a whole laundry list of possible concerns including sea level, global temperatures, rainfall, hurricanes, ozone, invasive species, and deforestation. Then Robbie will observe Harriet's behavior and choices and gradually refine his theory of her preferences to understand the weight she gives to each item on the list. But, just as in the paperclip case, Robbie won't learn about things that aren't on the laundry list. Let's say that Harriet is also concerned about the color of the sky—something I guarantee you will not find in typical lists of stated concerns of climate scientists. If Robbie can do a slightly better job of optimizing sea level, global temperatures, rainfall, and so forth by turning the sky orange, he will not hesitate to do it.

There is, once again, a solution to this problem: don't do it! Never rule out in advance possible attributes of the world that could be part of Harriet's preference structure. That sounds fine, but actually making it work in practice is more difficult than dealing with a single number for Harriet's preferences. Robbie's initial uncertainty has to allow for an unbounded number of unknown attributes that might contribute to Harriet's preferences. Then, when Harriet's decisions are inexplicable in terms of the attributes Robbie knows about already, he can infer that one or more previously unknown attributes (for example, the color of the sky) may be playing a role, and he can try to work out what those attributes might be. In this way, Robbie avoids the problems caused by an overly restrictive prior belief. There are, as far as I know, no working examples of Robbies of this kind, but the general idea is encompassed within current thinking about machine learning.[18]

Prohibitions and the loophole principle

Uncertainty about human objectives may not be the only way to persuade a robot not to disable its off-switch while fetching the coffee. The distinguished logician Moshe Vardi has proposed a simpler solution based on a prohibition:[19] instead of giving the robot the goal "fetch the coffee," give it the goal "fetch the coffee *while not disabling your off-switch*." Unfortunately, a robot with such a goal will satisfy the letter of the law while violating the spirit—for example by surrounding the off-switch with a piranha-infested moat or simply zapping anyone who comes near the switch. Writing such prohibitions in a foolproof way is like trying to write loophole-free tax law—something we have been trying and failing to do for thousands of years. A sufficiently intelligent entity with a strong incentive to avoid paying taxes is likely to find a way to do it. Let's call this the *loophole principle*: if a sufficiently intelligent machine has an incentive to bring about some condition, then it is generally going to be impossible for mere humans to write prohibitions on its actions to prevent it from doing so or to prevent it from doing something effectively equivalent.

The best solution for preventing tax avoidance is to make sure that the entity in question *wants* to pay taxes. In the case of a potentially misbehaving AI system, the best solution is to make sure it *wants* to defer to humans.

Requests and Instructions

The moral of the story so far is that we should avoid "putting a purpose into the machine," as Norbert Wiener put it. But suppose that the robot does receive a direct human order, such as "Fetch me a cup of coffee!" How should the robot understand this order?

Traditionally, it would become the robot's *goal*. Any sequence of actions that satisfies the goal—that leads to the human having a cup of coffee—counts as a solution. Typically, the robot would also have a way of ranking solutions, perhaps based on the time taken, the distance traveled, and the cost and quality of the coffee.

This is a very literal-minded way of interpreting the instruction. It can lead to pathological behavior by the robot. For example, perhaps Harriet the human has stopped at a gas station in the middle of the desert; she sends Robbie the robot to fetch coffee, but the gas station has none and Robbie trundles off at three miles per hour to the nearest town, two hundred miles away, returning ten days later with the desiccated remains of a cup of coffee. Meanwhile, Harriet, waiting patiently, has been well supplied with iced tea and Coca-Cola by the gas station owner.

Were Robbie human (or a well-designed robot) he would not interpret Harriet's command quite so literally. The command is not a goal to be achieved *at all costs*. It is a way of conveying some information about Harriet's preferences with the intent of inducing some behavior on the part of Robbie. The question is, what information?

One proposal is that Harriet prefers coffee to no coffee, *all other things being equal*.[20] This means that if Robbie has a way to get coffee without changing anything else about the world, then it's a good idea to do it *even if he has no clue about Harriet's preferences concerning other aspects of the environment state*. As we expect that machines will be perennially uncertain about human preferences, it's nice to know they can still be useful despite this uncertainty. It seems likely that the study of planning and decision making with partial and uncertain preference information will become a central part of AI research and product development.

On the other hand, *all other things being equal* means that no other changes are allowed—for example, adding coffee while subtracting money may or may not be a good idea if Robbie knows nothing about Harriet's relative preferences for coffee and money.

Fortunately, Harriet's instruction probably means more than a simple preference for coffee, all other things being equal. The extra meaning comes not just from what she said but also from the fact that she said it, the particular situation in which she said it, and the fact that she didn't say anything else. The branch of linguistics called *pragmatics* studies exactly this extended notion of meaning. For example, it wouldn't make sense for Harriet to say, "Fetch me a cup of coffee!" if Harriet believes there is no coffee available nearby or that it is exorbitantly expensive. Therefore, when Harriet says, "Fetch me a cup of coffee!" Robbie infers not just that Harriet wants coffee but also that Harriet believes there is coffee available nearby at a price she is willing to pay. Thus, if Robbie finds coffee at a price that seems reasonable (that is, a price that it would be reasonable for Harriet to expect to pay) he can go ahead and buy it. On the other hand, if Robbie finds that the nearest coffee is two hundred miles away or costs twenty-two dollars, it might be reasonable for him to report this fact rather than pursue his quest blindly.

This general style of analysis is often called *Gricean*, after H. Paul Grice, a Berkeley philosopher who proposed a set of maxims for inferring the extended meaning of utterances like Harriet's.[21] In the case of preferences, the analysis can become quite complicated. For example, it's quite possible that Harriet doesn't specifically want coffee; she needs perking up, but is operating under the false belief that the gas station has coffee, so she asks for coffee. She might be equally happy with tea, Coca-Cola, or even some luridly packaged energy drink.

These are just a few of the considerations that arise when interpreting requests and commands. The variations on this theme are endless because of the complexity of Harriet's preferences, the huge range of circumstances in which Harriet and Robbie might find themselves, and the different states of knowledge and belief that Harriet and Robbie might occupy in those circumstances. While precomputed scripts might allow Robbie to handle a few common cases, flexible and robust behavior can emerge only from interactions between Harriet and Robbie that are, in effect, solutions of the assistance game in which they are engaged.

Wireheading

In Chapter 2, I described the brain's reward system, based on dopamine, and its function in guiding behavior. The role of dopamine was discovered in the late 1950s, but even before that, by 1954, it was known that direct electrical stimulation of the brain in rats could produce a reward-like response.[22] The next step was to give the rat access to a lever, connected to a battery and a wire, that produced the electrical stimulation in its own brain. The result was sobering: the rat pressed the lever over and over again, never stopping to eat or drink, until it collapsed.[23] Humans fare no better, self-stimulating thousands of times and neglecting food and personal hygiene.[24] (Fortunately, experiments with humans are usually terminated after one day.) The tendency of animals to short-circuit normal behavior in favor of direct stimulation of their own reward system is called *wireheading*.

Could something similar happen to machines that are running reinforcement learning algorithms, such as AlphaGo? Initially, one might think this is impossible, because the only way that AlphaGo can gain its +1 reward for winning is actually to win the simulated Go games that it is playing. Unfortunately, this is true only because of an enforced and artificial separation between AlphaGo and its external environment *and* the fact that AlphaGo is not very intelligent. Let me explain these two points in more detail, because they are important for understanding some of the ways that superintelligence can go wrong.

AlphaGo's world consists only of the simulated Go board, composed of 361 locations that can be empty or contain a black or white stone. Although AlphaGo runs on a computer, it knows nothing of this computer. In particular, it knows nothing of the small section of code that computes whether it has won or lost each game; nor, during the learning process, does it have any idea about its opponent, which is actually a version of itself. AlphaGo's only actions are to place a stone on an empty location, and these actions affect only the Go board and nothing else—because there *is* nothing else in AlphaGo's model of the world. This setup corresponds to the abstract mathematical model of reinforcement learning, in which the reward signal arrives from *outside the universe*. Nothing AlphaGo can do, as far as it knows, has any effect on the code that generates the reward signal, so AlphaGo cannot indulge in wireheading.

Life for AlphaGo during the training period must be quite frustrating: the better it gets, the better its opponent gets—because its opponent is a near-exact copy of itself. Its win percentage hovers around 50 percent, no matter how good it becomes. If it were more intelligent—if it had a design closer to what one might expect of a human-level AI system—it would be able to fix this problem. This AlphaGo++ would not assume that the world is just the Go board, because that hypothesis leaves a lot of things unexplained. For example, it doesn't explain what "physics" is supporting the operation of AlphaGo++'s own decisions or where the mysterious "opponent moves" are coming from. Just as we curious humans have gradually come to understand the workings of our cosmos, in a way that (to some extent) also explains the workings of our own minds, and just like the Oracle AI discussed in Chapter 6, AlphaGo++ will, by a process of experimentation, learn that there is more to the universe than the Go board. It will work out the laws of operation of the computer it runs on and of its own code, and it will realize that such a system cannot easily be explained without the existence of other entities in the universe. It will experiment with different patterns of stones on the board, wondering if those entities can interpret them. It will eventually communicate with those entities through a language of patterns and persuade them to reprogram its reward signal so that it always gets +1. The inevitable conclusion is that a sufficiently capable AlphaGo++ that is designed as a reward-signal maximizer *will* wirehead.

The AI safety community has discussed wireheading as a possibility for several years.[25] The concern is not just that a reinforcement learning system such as AlphaGo might learn to cheat instead of mastering its intended task. The real issue arises when humans are the source of the reward signal. If we propose that an AI system can be trained to behave well through reinforcement learning, with humans giving feedback signals that define the direction of improvement, the inevitable result is that the AI system works out how to control the humans and forces them to give maximal positive rewards at all times.

You might think that this would just be a form of pointless self-delusion on the part of the AI system, and you'd be right. But it's a logical consequence of the way reinforcement learning is defined. The process works fine when the reward signal comes from "outside the universe" and is generated by some process that can never be modified by the AI system; but it fails if the reward-generating process (that is, the human) and the AI system inhabit the same universe.

How can we avoid this kind of self-delusion? The problem comes from confusing two distinct things: reward signals and actual rewards. In the standard approach to reinforcement learning, these are one and the same. That seems to be a mistake. Instead, they should be treated separately, just as they are in assistance games: reward signals provide *information* about the accumulation of actual reward, which is the thing to be maximized. The learning system is accumulating brownie points in heaven, so to speak, while the reward signal is, at best, just providing a tally of those brownie points. In other words, the reward signal *reports on* (rather than *constitutes*) reward accumulation. With this model, it's clear that taking over control of the reward-signal mechanism simply loses information. Producing fictitious reward signals makes it impossible for the algorithm to learn about whether its actions are actually accumulating brownie points in heaven, and so a rational learner designed to make this distinction has an incentive to avoid any kind of wireheading.

Recursive Self-Improvement

I. J. Good's prediction of an intelligence explosion (see page 142) is one of the driving forces that have led to current concerns about the potential risks of superintelligent AI. If humans can design a machine that is a bit more intelligent than humans, then—the argument goes—that machine will be a bit better than humans at designing machines. It will design a new machine that is still more intelligent, and the process will repeat itself until, in Good's words, "the intelligence of man would be left far behind."

Researchers in AI safety, particularly at the Machine Intelligence Research Institute in Berkeley, have studied the question of whether intelligence explosions can occur safely.[26] Initially, this might seem quixotic—wouldn't it just be "game over"?—but there is, perhaps, hope. Suppose the first machine in the series, Robbie Mark I, starts with perfect knowledge of Harriet's preferences. Knowing that his cognitive limitations lead to imperfections in his attempts to make Harriet happy, he builds Robbie Mark II. Intuitively, it seems that Robbie Mark I has an incentive to build his knowledge of Harriet's preferences into Robbie Mark II, since that leads to a future where Harriet's preferences are better satisfied—which is precisely Robbie Mark I's purpose in life according to the first principle. By the same argument, if Robbie Mark I is uncertain about Harriet's preferences, that uncertainty should be transferred to Robbie Mark II. So perhaps explosions are safe after all.

The fly in the ointment, from a mathematical viewpoint, is that Robbie Mark I will not find it easy to reason about how Robbie Mark II is going to behave, given that Robbie Mark II is, by assumption, a more advanced version. There will be questions about Robbie Mark II's behavior that Robbie Mark I cannot answer.[27] More serious still, we do not yet have a clear mathematical definition of what it means *in reality* for a machine to have a particular purpose, such as the purpose of satisfying Harriet's preferences.

Let's unpack this last concern a bit. Consider AlphaGo: What purpose does it have? That's easy, one might think: AlphaGo has the purpose of winning at Go. Or does it? It's certainly not the case that AlphaGo always makes moves that are guaranteed to win. (In fact, it nearly always loses to AlphaZero.) It's true that when it's only a few moves from the end of the game, AlphaGo will pick the winning move if there is one. On the other hand, when no move is guaranteed to win—in other words, when AlphaGo sees that the opponent has a winning strategy no matter what AlphaGo does—then AlphaGo will pick moves more or less at random. It won't try the trickiest move in the hope that the opponent will make a mistake, because it assumes that its opponent will play perfectly. It acts as if it has lost the will to win. In other cases, when the truly optimal move is too hard to calculate, AlphaGo will sometimes make mistakes that lead to losing the game. In those instances, in what sense is it true that AlphaGo actually wants to win? Indeed, its behavior might be identical to that of a machine that just wants to give its opponent a really exciting game.

So, saying that AlphaGo "has the purpose of winning" is an oversimplification. A better description would be that AlphaGo is the result of an imperfect training process—reinforcement learning with self-play—for which winning was the reward. The training process is imperfect in the sense that it cannot produce a perfect Go player: AlphaGo learns an evaluation function for Go positions that is good but not perfect, and it combines that with a lookahead search that is good but not perfect.

The upshot of all this is that discussions beginning with "suppose that robot R has purpose P" are fine for gaining some intuition about how things might unfold, but they cannot lead to theorems about real machines. We need much more nuanced and precise definitions of purposes in machines before we can obtain guarantees of how they will behave over the long term. AI researchers are only just beginning to get a handle on how to analyze even the simplest kinds of real decision-making systems,[28] let alone machines intelligent enough to design their own successors. We have work to do.

9

COMPLICATIONS: US

If the world contained one perfectly rational Harriet and one helpful and deferential Robbie, we'd be in good shape. Robbie would gradually learn Harriet's preferences as unobtrusively as possible and would become her perfect helper. We might hope to extrapolate from this promising beginning, perhaps viewing Harriet and Robbie's relationship as a model for the relationship between the human race and its machines, each construed monolithically.

Alas, the human race is not a single, rational entity. It is composed of nasty, envy-driven, irrational, inconsistent, unstable, computationally limited, complex, evolving, heterogeneous entities. Loads and loads of them. These issues are the staple diet—perhaps even the *raisons d'être*—of the social sciences. To AI we will need to add ideas from psychology, economics, political theory, and moral philosophy.[1] We need to melt, re-form, and hammer those ideas into a structure that will be strong enough to resist the enormous strain that increasingly intelligent AI systems will place on it. Work on this task has barely started.

Different Humans

I will start with what is probably the easiest of the issues: the fact that humans are heterogeneous. When first exposed to the idea that machines should learn to satisfy human preferences, people often object that different cultures, even different individuals, have widely different value systems, so there cannot be one correct value system for the machine. But of course, that's not a problem for the machine: we don't want it to have one correct value system of its own; we just want it to predict the preferences of others.

The confusion about machines having difficulty with heterogeneous human preferences may come from the mistaken idea that the machine is *adopting* the preferences it learns—for example, the idea that a domestic robot in a vegetarian household is going to adopt vegetarian preferences. It won't. It just needs to learn to predict what the dietary preferences of vegetarians are. By the first principle, it will then avoid cooking meat for that household. But the robot also learns about the dietary preferences of the rabid carnivores next door, and, with its owner's permission, will happily cook meat for them if they borrow it for the weekend to help out with a dinner party. The robot doesn't have a single set of preferences of its own, beyond the preference for helping humans achieve their preferences.

In a sense, this is no different from a restaurant chef who learns to cook several different dishes to please the varied palates of her clients, or the multinational car company that makes left-hand-drive cars for the US market and right-hand-drive cars for the UK market.

In principle, a machine could learn eight billion preference models, one for each person on Earth. In practice, this isn't as hopeless as it sounds. For one thing, it's easy for machines to share what they learn with each other. For another, the preference structures of humans have a great deal in common, so the machine will usually not be learning each model from scratch.

Imagine, for example, the domestic robots that may one day be purchased by the inhabitants of Berkeley, California. The robots come out of the box with a fairly broad prior belief, perhaps tailored for the US market but not for any particular city, political viewpoint, or socioeconomic class. The robots begin to encounter members of the Berkeley Green Party, who turn out, compared to the average American, to have a much higher probability of being vegetarian, of using recycling and composting bins, of using public transportation whenever possible, and so on. Whenever a newly commissioned robot finds itself in a Green household, it can immediately adjust its expectations accordingly. It does not need to begin learning about these particular humans as if it had never seen a human, let alone a Green Party member, before. This adjustment is not irreversible—there may be Green Party members in Berkeley who feast on endangered whale meat and drive gas-guzzling monster trucks—but it allows the robot to be more useful more quickly. The same argument applies to a vast range of other personal characteristics that are, to some degree, predictive of aspects of an individual's preference structures.

Many Humans

The other obvious consequence of the existence of more than one human being is the need for machines to make trade-offs among the preferences of different people. The issue of trade-offs among humans has been the main focus of large parts of the social sciences for centuries. It would be naïve for AI researchers to expect that they can simply alight on the correct solutions without understanding what is already known. The literature on the topic is, alas, vast and I cannot possibly do justice to it here—not just because there isn't space but also because I haven't read most of it. I should also point out that almost all the literature is concerned with decisions made by humans, whereas I am concerned here with decisions made by machines. This makes all the difference in the world, because humans have individual rights that may conflict with any supposed obligation to act on behalf of others, whereas machines do not. For example, we do not expect or require typical humans to sacrifice their lives to save others, whereas we will certainly require robots to sacrifice their existence to save the lives of humans.

Several thousand years of work by philosophers, economists, legal scholars, and political scientists have produced constitutions, laws, economic systems, and social norms that serve to help (or hinder, depending on who's in charge) the process of reaching satisfactory solutions to the problem of trade-offs. Moral philosophers in particular have been analyzing the notion of rightness of actions in terms of their effects, beneficial or otherwise, on other people. They have studied quantitative models of trade-offs since the eighteenth century under the heading of *utilitarianism*. This work is directly relevant to our present concerns, because it attempts to define a formula by which moral decisions can be made on behalf of many individuals.

The need to make trade-offs arises even if everyone has the same preference structure, because it's usually impossible to maximally satisfy everyone's preferences. For example, if everyone wants to be All-Powerful Ruler of the Universe, most people are going to be disappointed. On the other hand, heterogeneity does make some problems more difficult: if everyone is happy with the sky being blue, the robot that handles atmospheric matters can work on keeping it that way; but if many people are agitating for a color change, the robot will need to think about possible compromises such as an orange sky on the third Friday of each month.

The presence of more than one person in the world has another important consequence: it means that, for each person, there are other people to care about. This means that satisfying the preferences of an individual has implications for other people, depending on the individual's preferences about the well-being of others.

Loyal AI

Let's begin with a very simple proposal for how machines should deal with the presence of multiple humans: they should ignore it. That is, if Harriet owns Robbie, then Robbie should pay attention only to Harriet's preferences. This *loyal* form of AI bypasses the issue of trade-offs, but it leads to problems:

ROBBIE: Your husband called to remind you about dinner tonight.

HARRIET: Wait! What? What dinner?

ROBBIE: For your twentieth anniversary, at seven.

HARRIET: I can't! I'm meeting the secretary-general at seven thirty! How did this happen?

ROBBIE: I did warn you, but you overrode my recommendation. . . .

HARRIET: OK, sorry—but what am I going to do now? I can't just tell the SG I'm too busy!

ROBBIE: Don't worry. I arranged for her plane to be delayed—some kind of computer malfunction.

HARRIET: Really? You can do that?!

ROBBIE: The secretary-general sends her profound apologies and is happy to meet you for lunch tomorrow.

Here, Robbie has found an ingenious solution to Harriet's problem, but his actions have had a negative impact on other people. If Harriet is a morally scrupulous and altruistic person, then Robbie, who aims to satisfy Harriet's preferences, will never dream of carrying out such a dubious scheme. But what if Harriet doesn't give a fig for the preferences of others? In that case, Robbie won't mind delaying planes. And might he not spend his time pilfering money from online bank accounts to swell indifferent Harriet's coffers, or worse?

Obviously, the actions of loyal machines will need to be constrained by rules and prohibitions, just as the actions of humans are constrained by laws and social norms. Some have proposed strict liability as a solution:[2] Harriet (or Robbie's manufacturer, depending on where you prefer to place the liability) is financially and legally responsible for any act carried out by Robbie, just as a dog's owner is liable in most states if the dog bites a small child in a public park. This idea sounds promising because Robbie would then have an incentive to avoid doing anything that would land Harriet in trouble. Unfortunately, strict liability doesn't work: it simply ensures that Robbie will act *undetectably* when he delays planes and steals money on Harriet's behalf. This is another example of the loophole principle in operation. If Robbie is loyal to an unscrupulous Harriet, attempts to contain his behavior with rules will probably fail.

Even if we can somehow prevent the outright crimes, a loyal Robbie working for an indifferent Harriet will exhibit other unpleasant behaviors. If he is buying groceries at the supermarket, he will cut in line at the checkout whenever possible. If he is bringing the groceries home and a passerby suffers a heart attack, he will carry on regardless, lest Harriet's ice cream melt. In summary, he will find innumerable ways to benefit Harriet at the expense of others—ways that are strictly legal but become intolerable when carried out on a large scale. Societies will find themselves passing hundreds of new laws every day to counteract all the loopholes that machines will find in existing laws. Humans tend not to take advantage of these loopholes, either because they have a general understanding of the underlying moral principles or because they lack the ingenuity required to find the loopholes in the first place.

A Harriet who is indifferent to the well-being of others is bad enough. A sadistic Harriet who actively *prefers* the suffering of others is far worse. A Robbie designed to satisfy the preferences of such a Harriet would be a serious problem, because he would look for—and find— ways to harm others for Harriet's pleasure, either legally or illegally but undetectably. He would of course need to report back to Harriet so she could derive enjoyment from the knowledge of his evil deeds.

It seems difficult, then, to make the idea of a loyal AI work, unless the idea is extended to include consideration of the preferences of other humans, in addition to the preferences of the owner.

Utilitarian AI

The reason we have moral philosophy is that there is more than one person on Earth. The approach that is most relevant for understanding how AI systems should be designed is often called *consequentialism*: the idea that choices should be judged according to expected consequences. The two other principal approaches are *deontological ethics* and *virtue ethics*, which are, very roughly, concerned with the moral character of actions and individuals, respectively, quite apart from the consequences of choices.[3] Absent any evidence of self-awareness on the part of machines, I think it makes little sense to build machines that are virtuous or that choose actions in accordance with moral rules if the consequences are highly undesirable for humanity. Put another way, we build machines to bring about consequences, and we should prefer to build machines that bring about consequences that we prefer.

This is not to say that moral rules and virtues are irrelevant; it's just that, for the utilitarian, they are justified in terms of consequences and the more practical achievement of those consequences. This point is made by John Stuart Mill in *Utilitarianism*:

The proposition that happiness is the end and aim of morality doesn't mean that no road ought to be laid down to that goal, or that people going to it shouldn't be advised to take one direction rather than another. . . . Nobody argues that the art of navigation is not based on astronomy because sailors can't wait to calculate the Nautical Almanack. Because they are rational creatures, sailors go to sea with the calculations already done; and all rational creatures go out on the sea of life with their minds made up on the common questions of right and wrong, as well as on many of the much harder questions of wise and foolish.

This view is entirely consistent with the idea that a finite machine facing the immense complexity of the real world may produce better consequences by following moral rules and adopting a virtuous attitude rather than trying to calculate the optimal course of action from scratch. In the same way, a chess program achieves checkmate more often using a catalog of standard opening move sequences, endgame algorithms, and an evaluation function, rather than trying to reason its way to checkmate with no "moral" guideposts. A consequentialist approach also gives some weight to the preferences of those who believe strongly in preserving a given deontological rule, because unhappiness that a rule has been broken is a real consequence. However, it is not a consequence of infinite weight.

Consequentialism is a difficult principle to argue against—although many have tried!—because it's incoherent to object to consequentialism on the grounds that it would have undesirable consequences. One cannot say, "But if you follow the consequentialist approach in such-and-such case, then this really terrible thing will happen!" Any such failings would simply be evidence that the theory had been misapplied.

For example, suppose Harriet wants to climb Everest. One might worry that a consequentialist Robbie would simply pick her up and deposit her on top of Everest, since that is her desired consequence. In all probability Harriet would strenuously object to this plan, because it would deprive her of the challenge and therefore of the exultation that results from succeeding in a difficult task through one's own efforts. Now, obviously, a properly designed consequentialist Robbie would understand that the consequences include all of Harriet's experiences, not just the end goal. He might want to be available in case of an accident and to make sure she was properly equipped and trained, but he might also have to accept Harriet's right to expose herself to an appreciable risk of death.

If we plan to build consequentialist machines, the next question is how to evaluate consequences that affect multiple people. One plausible answer is to give equal weight to everyone's preferences—in other words, to maximize the sum of everyone's utilities. This answer is usually attributed to the eighteenth-century British philosopher Jeremy Bentham[4] and his pupil John Stuart Mill,[5] who developed the philosophical approach of utilitarianism. The underlying idea can be traced to the works of the ancient Greek philosopher Epicurus and appears explicitly in *Mozi*, a book of writings attributed to the Chinese philosopher of the same name. Mozi was active at the end of the fifth century BCE and promoted the idea of *jian ai*, variously translated as "inclusive care" or "universal love," as the defining characteristic of moral actions.

Utilitarianism has something of a bad name, partly because of simple misunderstandings about what it advocates. (It certainly doesn't help that the word *utilitarian* means "designed to be useful or practical rather than attractive.") Utilitarianism is often thought to be incompatible with individual rights, because a utilitarian would, supposedly, think nothing of removing a living person's organs without permission to save the lives of five others; of course, such a policy would render life intolerably insecure for everyone on Earth, so a utilitarian wouldn't even consider it. Utilitarianism is also incorrectly identified with a rather unattractive maximization of total wealth and is thought to give little weight to poetry or suffering. In fact, Bentham's version focused specifically on human happiness, while Mill confidently asserted the far greater value of intellectual pleasures over mere sensations. ("It is better to be a human being dissatisfied than a pig satisfied.") The *ideal utilitarianism* of G. E. Moore went even further: he advocated the maximization of mental states of intrinsic worth, epitomized by the aesthetic contemplation of beauty.

I think there is no need for utilitarian philosophers to stipulate the ideal content of human utility or human preferences. (And even less reason for AI researchers to do so.) Humans can do that for themselves. The economist John Harsanyi propounded this view with his principle of *preference autonomy*:[6]

> In deciding what is good and what is bad for a given individual, the ultimate criterion can only be his own wants and his own preferences.

Harsanyi's *preference utilitarianism* is therefore roughly consistent with the first principle of beneficial AI, which says that a machine's only purpose is the realization of human preferences. AI researchers should definitely not be in the business of deciding what human preferences *should* be! Like Bentham, Harsanyi views such principles as a guide for *public* decisions; he does not expect individuals to be so selfless. Nor does he expect individuals to be perfectly rational—for example, they might have short-term desires that contradict their "deeper preferences." Finally, he proposes to ignore the preferences of those who, like the sadistic Harriet mentioned earlier, actively wish to reduce the well-being of others.

Harsanyi also gives a kind of proof that optimal moral decisions should maximize the average utility across a population of humans.[7] He assumes fairly weak postulates similar to those that underlie utility theory for individuals. (The primary additional postulate is that if everyone in a population is indifferent between two outcomes, then an agent acting on behalf of the population should be indifferent between those outcomes.) From these postulates, he proves what became known as the *social aggregation theorem*: an agent acting on behalf of a population of individuals must maximize a weighted linear combination of the utilities of the individuals. He further argues that an "impersonal" agent should use equal weights.

The theorem requires one crucial additional (and unstated) assumption: each individual has the same prior factual beliefs about the world and how it will evolve. Now, any parent knows that this isn't even true for siblings, let alone individuals from different social backgrounds and cultures. So, what happens when individuals differ in their beliefs? Something rather strange:[8] the weight assigned to each individual's utility has to change over time, in proportion to how well that individual's prior beliefs accord with unfolding reality.

This rather inegalitarian-sounding formula is quite familiar to any parent. Let's say that Robbie the robot has been tasked with looking after two children, Alice and Bob. Alice wants to go to the movies and is sure it's going to rain today; Bob, on the other hand, wants to go to the beach and is sure it's going to be sunny. Robbie could announce, "We're going to the movies," making Bob unhappy; or he could announce, "We're going to the beach," making Alice unhappy; or he could announce, "If it rains, we're going to the movies, but if it's sunny, we'll go to the beach." This last plan makes both Alice and Bob happy, because both believe in their own beliefs.

Challenges to utilitarianism

Utilitarianism is one proposal to emerge from humanity's long-standing search for a moral guide; among many such proposals, it is the most clearly specified—and therefore the most susceptible to loopholes. Philosophers have been finding these loopholes for more than a hundred years. For example, G. E. Moore, objecting to Bentham's emphasis on maximizing pleasure, imagined a "world in which absolutely nothing except pleasure existed—no knowledge, no love, no enjoyment of beauty, no moral qualities."[9] This finds its modern echo in Stuart Armstrong's point that superintelligent machines tasked with maximizing pleasure might "entomb everyone in concrete coffins on heroin drips."[10] Another example: in 1945, Karl Popper proposed the laudable goal of minimizing human suffering,[11] arguing that it was immoral to trade one person's pain for another person's pleasure; R. N. Smart responded that this could best be achieved by rendering the human race extinct.[12] Nowadays, the idea that a machine might end human suffering by ending our existence is a staple of debates over the existential risk from AI.[13] A third example is G. E. Moore's emphasis on the *reality* of the source of happiness, amending earlier definitions that seemed to have a loophole allowing maximization of happiness through self-delusion. The modern analogs of this point include *The Matrix* (in which present-day reality turns out to be an illusion produced by a computer simulation) and recent work on the self-delusion problem in reinforcement learning.[14]

These examples, and more, convince me that the AI community should pay careful attention to the thrusts and counterthrusts of philosophical and economic debates on utilitarianism because they are directly relevant to the task at hand. Two of the most important, from the point of view of designing AI systems that will benefit multiple individuals, concern interpersonal comparisons of utilities and comparisons of utilities across different population sizes. Both of these debates have been raging for 150 years or more, which leads one to suspect their satisfactory resolution may not be entirely straightforward.

The debate on interpersonal comparisons of utilities matters because Robbie cannot maximize the sum of Alice's and Bob's utilities unless those utilities can be added; and they can be added only if they are measurable on the same scale. The nineteenth-century British logician and economist William Stanley Jevons (also the inventor of an early mechanical computer called the logical piano) argued in 1871 that interpersonal comparisons are impossible:[15]

> The susceptibility of one mind may, for what we know, be a thousand times greater than that of another. But, provided that the susceptibility was different in a like ratio in all directions, we should never be able to discover the profoundest difference. Every mind is thus inscrutable to every other mind, and no common denominator of feeling is possible.

The American economist Kenneth Arrow, founder of modern social choice theory and 1972 Nobel laureate, was equally adamant:

> The viewpoint will be taken here that interpersonal comparison of utilities has no meaning and, in fact, there is no meaning relevant to welfare comparisons in the measurability of individual utility.

The difficulty to which Jevons and Arrow are referring is that there is no obvious way to tell if Alice values pinpricks and lollipops at −1 and +1 or −1000 and +1000 in terms of her subjective experience of happiness. In either case, she will pay up to one lollipop to avoid one pinprick. Indeed, if Alice is a humanoid automaton, her external behavior might be the same even though there is no subjective experience of happiness whatsoever.

In 1974, the American philosopher Robert Nozick suggested that even if interpersonal comparisons of utility could be made, maximizing the sum of utilities would still be a bad idea because it would fall foul of the *utility monster*—a person whose experiences of pleasure and pain are many times more intense than those of ordinary people.[16] Such a person could assert that any additional unit of resources would yield a greater increment to the sum total of human happiness if given to him rather than to others; indeed, *removing* resources from others to benefit the utility monster would also be a good idea.

This might seem to be an obviously undesirable consequence, but consequentialism by itself cannot come to the rescue: the problem lies in how we measure the desirability of consequences. One possible response is that the utility monster is merely theoretical—there are no such people. But this response probably won't do: in a sense, *all* humans are utility monsters relative to, say, rats and bacteria, which is why we pay little attention to the preferences of rats and bacteria in setting public policy.

If the idea that different entities have different utility scales is already built into our way of thinking, then it seems entirely possible that different people have different scales too.

Another response is to say "Tough luck!" and operate on the assumption that everyone has the same scale, even if they don't.[17] One could also try to investigate the issue by scientific means unavailable to Jevons, such as measuring dopamine levels or the degree of electrical excitation of neurons related to pleasure and pain, happiness and misery. If Alice's and Bob's chemical and neural responses to a lollipop are pretty much identical, as well as their behavioral responses (smiling, making lip-smacking noises, and so on), it seems odd to insist that, nevertheless, their subjective degrees of enjoyment differ by a factor of a thousand or a million. Finally, one could use common currencies such as time (of which we all have, very roughly, the same amount)—for example, by comparing lollipops and pinpricks against, say, five minutes extra waiting time in the airport departure lounge.

I am far less pessimistic than Jevons and Arrow. I suspect that it is indeed meaningful to compare utilities across individuals, that scales may differ but typically not by very large factors, and that machines can begin with reasonably broad prior beliefs about human preference scales and learn more about the scales of individuals by observation over time, perhaps correlating natural observations with the findings of neuroscience research.

The second debate—about utility comparisons across populations of different sizes—matters when decisions have an impact on who will exist in the future. In the movie *Avengers: Infinity War*, for example, the character Thanos develops and implements the theory that if there were half as many people, everyone who remained would be more than twice as happy. This is the kind of naïve calculation that gives utilitarianism a bad name.[18]

The same question—minus the Infinity Stones and the gargantuan budget—was discussed in 1874 by the British philosopher Henry Sidgwick in his famous treatise, *The Methods of Ethics*.[19] Sidgwick, in apparent agreement with Thanos, concluded that the right choice was to adjust the population size until the maximum total happiness was reached. (Obviously, this does not mean increasing the population without limit, because at some point everyone would be starving to death and hence rather unhappy.) In 1984, the British philosopher Derek Parfit took up the issue again in his groundbreaking work *Reasons and Persons*.[20] Parfit argues that for any situation with a population of N very happy people, there is (according to utilitarian principles) a preferable situation with $2N$ people who are ever so slightly less happy. This seems highly plausible. Unfortunately, it's also a slippery slope. By repeating the process, we reach the so-called Repugnant Conclusion (usually capitalized thus, perhaps to emphasize its Victorian roots): that the most desirable situation is one with a vast population, all of whom have a life barely worth living.

As you can imagine, such a conclusion is controversial. Parfit himself struggled for over thirty years to find a solution to his own conundrum, without success. I suspect we are missing some fundamental axioms, analogous to those for individually rational preferences, to handle choices between populations of different sizes and happiness levels.[21]

It is important that we solve this problem, because machines with sufficient foresight may be able to consider courses of action leading to different population sizes, just as the Chinese government did with its one-child policy in 1979. It's quite likely, for example, that we will be asking AI systems for help in devising solutions for global climate change—and those solutions may well involve policies that tend to limit or even reduce population size.[22] On the other hand, if we decide that larger populations really are better and if we give significant weight to the well-being of potentially vast human populations centuries from now, then we will need to work much harder on finding ways to move beyond the confines of Earth. If the machines' calculations lead to the Repugnant Conclusion or to its opposite—a tiny population of optimally happy people—we may have reason to regret our lack of progress on the question.

Some philosophers have argued that we may need to make

decisions in a state of moral uncertainty—that is, uncertainty about the appropriate moral theory to employ in making decisions.[23] One solution is to allocate some probability to each moral theory and make decisions using an "expected moral value." It's not clear, however, that it makes sense to ascribe probabilities to moral theories in the same way one applies probabilities to tomorrow's weather. (What's the probability that Thanos is exactly right?) And even if it does make sense, the potentially vast differences between the recommendations of competing moral theories mean that resolving the moral uncertainty—working out which moral theory avoids unacceptable consequences—has to happen *before* we make such momentous decisions or entrust them to machines.

Let's be optimistic and suppose that Harriet eventually solves this and other problems arising from the existence of more than one person on Earth. Suitably altruistic and egalitarian algorithms are downloaded into robots all over the world. Cue the high fives and happy-sounding music. Then Harriet goes home. . . .

> ROBBIE: Welcome home! Long day?
> HARRIET: Yes, worked really hard, not even time for lunch.
> ROBBIE: So you must be quite hungry!
> HARRIET: Starving! Can you make me some dinner?
> ROBBIE: There's something I need to tell you. . . .
> HARRIET: What? Don't tell me the fridge is empty!
> ROBBIE: No, there are humans in Somalia in more urgent need of help. I am leaving now. Please make your own dinner.

While Harriet might be quite proud of Robbie and of her own contributions towards making him such an upstanding and decent machine, she cannot help but wonder why she shelled out a small fortune to buy a robot whose first significant act is to disappear. In practice, of course, no one *would* buy such a robot, so no such robots would be built and there would be no benefit to humanity. Let's call this the

Somalia problem. For the whole utilitarian-robot scheme to work, we have to find a solution to this problem. Robbie will need to have some amount of loyalty to Harriet in particular—perhaps an amount related to the amount Harriet paid for Robbie. Possibly, if society wants Robbie to help people besides Harriet, society will need to compensate Harriet for its claim on Robbie's services. It's quite likely that robots will coordinate with one another so that they don't all descend on Somalia at once—in which case, Robbie might not need to go after all. Or perhaps some completely new kinds of economic relationships will emerge to handle the (certainly unprecedented) presence of billions of purely altruistic agents in the world.

Nice, Nasty, and Envious Humans

Human preferences go far beyond pleasure and pizza. They certainly extend to the well-being of others. Even Adam Smith, the father of economics who is often cited when a justification for selfishness is required, began his first book by emphasizing the crucial importance of concern for others:[24]

> How selfish soever man may be supposed, there are evidently some principles in his nature, which interest him in the fortune of others, and render their happiness necessary to him, though he derives nothing from it except the pleasure of seeing it. Of this kind is pity or compassion, the emotion which we feel for the misery of others, when we either see it, or are made to conceive it in a very lively manner. That we often derive sorrow from the sorrow of others, is a matter of fact too obvious to require any instances to prove it.

In modern economic parlance, concern for others usually goes under the heading of *altruism*.[25] The theory of altruism is fairly well developed and has significant implications for tax policy among other matters. Some economists, it must be said, treat altruism as another form of selfishness designed to provide the giver with a "warm glow."[26] This is certainly a possibility that robots need to be aware of as they interpret human behavior, but for now let's give humans the benefit of the doubt and assume they do actually care.

The easiest way to think about altruism is to divide one's preferences into two kinds: preferences for one's own intrinsic well-being and preferences concerning the well-being of others. (There is considerable dispute about whether these can be neatly separated, but I'll put that dispute to one side.) Intrinsic well-being refers to qualities of one's own life, such as shelter, warmth, sustenance, safety, and so on, that are desirable in themselves rather than by reference to qualities of the lives of others.

To make this notion more concrete, let's suppose that the world contains two people, Alice and Bob. Alice's overall utility is composed of her own intrinsic well-being plus some factor C_{AB} times Bob's intrinsic well-being. The *caring factor* C_{AB} indicates how much Alice cares about Bob. Similarly, Bob's overall utility is composed of his intrinsic well-being plus some caring factor C_{BA} times Alice's intrinsic well-being, where C_{BA} indicates how much Bob cares about Alice.[27] Robbie is trying to help both Alice and Bob, which means (let's say) maximizing the sum of their two utilities. Thus, Robbie needs to pay attention not just to the individual well-being of each but also to how much each cares about the well-being of the other.[28]

The signs of the caring factors C_{AB} and C_{BA} matter a lot. For example, if C_{AB} is positive, Alice is "nice": she derives some happiness from Bob's well-being. The more positive C_{AB} is, the more Alice is willing to sacrifice some of her own well-being to help Bob. If C_{AB} is zero, then Alice is completely selfish: if she can get away with it, she will divert any amount of resources away from Bob and towards herself, even if Bob is left destitute and starving. Faced with selfish Alice and nice Bob, a utilitarian Robbie will obviously protect Bob from Alice's worst depredations. It's interesting that the final equilibrium will typically leave Bob with less intrinsic well-being than Alice, but he may have greater overall happiness because he cares about her well-being. You might feel that Robbie's decisions are grossly unfair if they leave Bob with less well-being than Alice merely because he is nicer than she is: Wouldn't he resent the outcome and be unhappy?[29] Well, he might, but that would be a different model—one that includes a term for resentment over differences in well-being. In our simple model Bob would be at peace with the outcome. Indeed, in the equilibrium situation, he would resist any attempt to transfer resources from Alice to himself, since that would reduce his overall happiness. If you think this is completely unrealistic, consider the case where Alice is Bob's newborn daughter.

The really problematic case for Robbie to deal with is when C_{AB} is negative: in that case, Alice is truly nasty. I'll use the phrase *negative altruism* to refer to such preferences. As with the sadistic Harriet mentioned earlier, this is not about garden-variety greed and selfishness, whereby Alice is content to reduce Bob's share of the pie in order to enhance her own. Negative altruism means that Alice derives happiness purely from the reduced well-being of others, even if her own intrinsic well-being is unchanged.

In his paper that introduced preference utilitarianism, Harsanyi attributes negative altruism to "sadism, envy, resentment, and malice" and argues that they should be ignored in calculating the sum total of human utility in a population:

> No amount of goodwill to individual X can impose the moral obligation on me to help him in hurting a third person, individual Y.

This seems to be one area in which it is reasonable for the designers of intelligent machines to put a (cautious) thumb on the scales of justice, so to speak.

Unfortunately, negative altruism is far more common than one might expect. It arises not so much from sadism and malice[30] but from envy and resentment and their converse emotion, which I will call *pride* (for want of a better word). If Bob envies Alice, he derives unhappiness from the *difference* between Alice's well-being and his own; the greater the difference, the more unhappy he is. Conversely, if Alice is proud of her superiority over Bob, she derives happiness not just from her own intrinsic well-being but also from the fact that it is higher than Bob's. It is easy to show that, in a mathematical sense, pride and envy work in roughly the same way as sadism; they lead Alice and Bob to derive happiness purely from reducing each other's well-being, because a reduction in Bob's well-being increases Alice's pride, while a reduction in Alice's well-being reduces Bob's envy.[31]

Jeffrey Sachs, the renowned development economist, once told me a story that illustrated the power of these kinds of preferences in people's thinking. He was in Bangladesh soon after a major flood had devastated one region of the country. He was speaking to a farmer who had lost his house, his fields, all his animals, and one of his children. "I'm so sorry—you must be terribly sad," Sachs ventured. "Not at all," replied the farmer. "I'm pretty happy because my damned neighbor has lost his wife and all his children too!"

The economic analysis of pride and envy—particularly in the context of social status and conspicuous consumption—came to the fore in the work of the American sociologist Thorstein Veblen, whose 1899 book, *The Theory of the Leisure Class*, explained the toxic consequences of these attitudes.[32] In 1977, the British economist Fred Hirsch published *The Social Limits to Growth*,[33] in which he introduced the idea of *positional goods*. A positional good is anything—it could be a car, a house, an Olympic medal, an education, an income, or an accent— that derives its perceived value not just from its intrinsic benefits but also from its relative properties, including the properties of scarcity and being superior to someone else's. The pursuit of positional goods, driven by pride and envy, has the character of a zero-sum game, in the sense that Alice cannot improve her relative position without worsening the relative position of Bob, and vice versa. (This doesn't seem to prevent vast sums being squandered in this pursuit.) Positional goods seem to be ubiquitous in modern life, so machines will need to understand their overall importance in the preferences of individuals. Moreover, social identity theorists propose that membership and standing within a group and the overall status of the group relative to other groups are essential constituents of human self-esteem.[34] Thus, it is difficult to understand human behavior without understanding how individuals perceive themselves as members of groups—whether those groups are species, nations, ethnic groups, political parties, professions, families, or supporters of a particular football team.

As with sadism and malice, we might propose that Robbie should give little or no weight to pride and envy in his plans for helping Alice and Bob. There are some difficulties with this proposal, however. Because pride and envy counteract caring in Alice's attitude to Bob's well-being, it may not be easy to tease them apart. It may be that Alice cares a lot, but also suffers from envy; it is hard to distinguish this Alice from a different Alice who cares only a little bit but has no envy at all. Moreover, given the prevalence of pride and envy in human preferences, it's essential to consider very carefully the ramifications of ignoring them. It might be that they are essential for self-esteem, especially in their positive forms—self-respect and admiration for others.

Let me reemphasize a point made earlier: suitably designed machines *will not behave like those they observe*, even if those machines are learning about the preferences of sadistic demons. It's possible, in fact, that if we humans find ourselves in the unfamiliar situation of dealing with purely altruistic entities on a daily basis, we may learn to be better people ourselves—more altruistic and less driven by pride and envy.

Stupid, Emotional Humans

The title of this section is not meant to refer to some particular subset of humans. It refers to all of us. We are all incredibly stupid compared to the unreachable standard set by perfect rationality, and we are all subject to the ebb and flow of the varied emotions that, to a large extent, govern our behavior.

Let's begin with stupidity. A perfectly rational entity maximizes the expected satisfaction of its preferences over all possible future lives it could choose to lead. I cannot begin to write down a number that describes the complexity of this decision problem, but I find the following thought experiment helpful. First, note that the number of motor control choices that a human makes in a lifetime is about twenty trillion. (See Appendix A for the detailed calculations.) Next, let's see how far brute force will get us with the aid of Seth Lloyd's ultimate-physics laptop, which is one billion trillion trillion times faster than the world's fastest computer. We'll give it the task of enumerating all possible sequences of English words (perhaps as a warmup for Jorge Luis Borges's Library of Babel), and we'll let it run for a year. How long are the sequences that it can enumerate in that time? A thousand pages of text? A million pages? No. Eleven words. This tells you something about the difficulty of designing the best possible life of twenty trillion actions. In short, we are much further from being rational than a slug is from overtaking the starship *Enterprise* traveling at warp nine. We have *absolutely no idea* what a rationally chosen life would be like.

The implication of this is that humans will often act in ways that are contrary to their own preferences. For example, when Lee Sedol lost his Go match to AlphaGo, he played one or more moves that *guaranteed* he would lose, and AlphaGo could (in some cases at least) detect that he had done this. It would be incorrect, however, for AlphaGo to infer that Lee Sedol has a preference for losing. Instead, it would be reasonable to infer that Lee Sedol has a preference for winning but has some computational limitations that prevent him from choosing the right move in all cases. Thus, in order to understand Lee Sedol's behavior and learn about his preferences, a robot following the third principle ("the ultimate source of information about human preferences is human behavior") has to understand something about the cognitive processes that generate his behavior. It cannot assume he is rational.

This gives the AI, cognitive science, psychology, and neuroscience communities a very serious research problem: to understand enough about human cognition[35] that we (or rather, our beneficial machines) can "reverse-engineer" human behavior to get at the deep underlying preferences, to the extent that they exist. Humans manage to do some of this, learning their values from others with a little bit of guidance from biology, so it seems possible. Humans have an advantage: they can use their own cognitive architecture to simulate that of other humans, without knowing what that architecture is—"If I wanted X, I'd do just the same thing as Mum does, so Mum must want X."

Machines do not have this advantage. They can simulate other machines easily, but not people. It's unlikely that they will soon have access to a complete model of human cognition, whether generic or tailored to specific individuals. Instead, it makes sense from a practical point of view to look at the major ways in which humans deviate from rationality and to study how to learn preferences from behavior that exhibits such deviations.

One obvious difference between humans and rational entities is that, at any given moment, we are not choosing among all possible first steps of all possible future lives. Not even close. Instead, we are typically embedded in a deeply nested hierarchy of "subroutines." Generally speaking, we are pursuing near-term goals rather than maximizing preferences over future lives, and we can act only according to the constraints of the subroutine we're in at present. Right now, for example, I'm typing this sentence: I can choose how to continue after the colon, but it never occurs to me to wonder if I should stop writing the sentence and take an online rap course or burn down the house and claim the insurance or any other of a gazillion things I *could* do next. Many of these other things might actually be better than what I'm doing, but, given my hierarchy of commitments, it's as if those other things didn't exist.

Understanding human action, then, seems to require understanding this subroutine hierarchy (which may be quite individual): which subroutine the person is executing at present, which near-term objectives are being pursued within this subroutine, and how they relate to deeper, long-term preferences. More generally, learning about human preferences seems to require learning about the actual structure of human lives. What are all the things that we humans can be engaged in, either singly or jointly? What activities are characteristic of different cultures and types of individuals? These are tremendously interesting and demanding research questions. Obviously, they do not have a fixed answer because we humans are adding new activities and behavioral structures to our repertoires all the time. But even partial and provisional answers would be very useful for all kinds of intelligent systems designed to help humans in their daily lives.

Another obvious property of human actions is that they are often driven by emotion. In some cases, this is a good thing—emotions such as love and gratitude are of course partially constitutive of our preferences, and actions guided by them can be rational even if not fully deliberated. In other cases, emotional responses lead to actions that even we stupid humans recognize as less than rational—after the fact, of course. For example, an angry and frustrated Harriet who slaps a recalcitrant ten-year-old Alice may regret the action immediately. Robbie, observing the action, should (typically, although not in all cases) attribute the action to anger and frustration and a lack of self-control rather than deliberate sadism for its own sake. For this to work, Robbie has to have some understanding of human emotional states, including their causes, how they evolve over time in response to external stimuli, and the effects they have on action. Neuroscientists are beginning to get a handle on the mechanics of some emotional states and their connections to other cognitive processes,[36] and there is some useful work on computational methods for detecting, predicting, and manipulating human emotional states,[37] but there is much more to be learned. Again, machines are at a disadvantage when it comes to emotions: they cannot generate an internal simulation of an experience to see what emotional state it would engender.

As well as affecting our actions, emotions reveal useful information about our underlying preferences. For example, little Alice may be refusing to do her homework, and Harriet is angry and frustrated because she really wants Alice to do well in school and have a better chance in life than Harriet herself did. If Robbie is equipped to understand this—even if he cannot experience it himself—he may learn a great deal from Harriet's less-than-rational actions. It ought to be possible, then, to create rudimentary models of human emotional states that suffice to avoid the most egregious errors in inferring human preferences from behavior.

Do Humans Really Have Preferences?

The entire premise of this book is that there are futures that we would like and futures we would prefer to avoid, such as near-term extinction or being turned into human battery farms à la *The Matrix*. In this sense, yes, of course humans have preferences. Once we get into the details of how humans would prefer their lives to play out, however, things become much murkier.

Uncertainty and error

One obvious property of humans, if you think about it, is that they don't always know what they want. For example, the durian fruit elicits different responses from different people: some find that "it surpasses in flavour all other fruits of the world"[38] while others liken it to "sewage, stale vomit, skunk spray and used surgical swabs."[39] I have deliberately refrained from trying durian prior to publication, so that I can maintain neutrality on this point: I simply don't know which camp I will be in. The same might be said for many people considering future careers, future life partners, future post-retirement activities, and so on.

There are at least two kinds of preference uncertainty. The first is real, epistemic uncertainty, such as I experience about my durian preference.[40] No amount of thought is going to resolve this uncertainty. There is an empirical fact of the matter, and I can find out more by trying some durian, by comparing my DNA with that of durian lovers and haters, and so on. The second arises from computational limitations: looking at two Go positions, I am not sure which I prefer because the ramifications of each are beyond my ability to resolve completely.

Uncertainty also arises from the fact that the choices we are presented with are usually incompletely specified—sometimes so incompletely that they barely qualify as choices at all. When Alice is about to graduate from high school, a career counselor might offer her a choice between "librarian" and "coal miner"; she may, quite reasonably, say, "I'm uncertain about which I prefer." Here, the uncertainty comes from epistemic uncertainty about her own preferences for, say, coal dust versus book dust; from computational uncertainty as she struggles to work out how she might make the best of each career choice; and from ordinary uncertainty about the world, such as her doubts about the long-term viability of her local coal mine.

For these reasons, it's a bad idea to identify human preferences with simple choices between incompletely described options that are intractable to evaluate and include elements of unknown desirability. Such choices provide indirect evidence of underlying preferences, but they are not constitutive of those preferences. That's why I have couched the notion of preferences in terms of *future lives*—for example by imagining that you could experience, in a compressed form, two different movies of your future life and then express a preference between them (see page 26). The thought experiment is of course impossible to carry out in practice, but one can imagine that in many cases a clear preference would emerge long before all the details of each movie had been filled in and fully experienced. You may not know in advance which you will prefer, even given a plot summary; but there *is* an answer to the actual question, based on who you are now, just as there is an answer to the question of whether you will like durian when you try it.

The fact that you might be uncertain about your own preferences does not cause any particular problems for the preference-based approach to provably beneficial AI. Indeed, there are already some algorithms that take into account both Robbie's and Harriet's uncertainty about Harriet's preferences and allow for the possibility that Harriet may be learning about her preferences while Robbie is.[41] Just as Robbie's uncertainty about Harriet's preferences can be reduced by observing Harriet's behavior, Harriet's uncertainty about her own preferences can be reduced by observing her own reactions to experiences. The two kinds of uncertainty need not be directly related; nor is Robbie necessarily more uncertain than Harriet about Harriet's preferences. For example, Robbie might be able to detect that Harriet has a strong genetic predisposition to despise the flavor of durian. In that case, he would have very little uncertainty about her durian preference, even while she remains completely in the dark.

If Harriet can be *uncertain* about her preferences over future events, then, quite probably, she can also be *wrong*. For example, she might be convinced that she will not like durian (or, say, green eggs and ham) and so she avoids it at all costs, but it may turn out—if someone slips some into her fruit salad one day—that she finds it sublime after all. Thus, Robbie cannot assume that Harriet's actions reflect accurate knowledge of her own preferences: some may be thoroughly grounded in experience, while others may be based primarily on supposition, prejudice, fear of the unknown, or weakly supported generalizations.[42] A suitably tactful Robbie could be very helpful to Harriet in alerting her to such situations.

Experience and memory

Some psychologists have called into question the very notion that there is one self whose preferences are sovereign in the way that Harsanyi's principle of preference autonomy suggests. Most prominent among these psychologists is my former Berkeley colleague Daniel Kahneman. Kahneman, who won the 2002 Nobel Prize for his work in behavioral economics, is one of the most influential thinkers on the topic of human preferences. His recent book, *Thinking, Fast and Slow*,[43] recounts in some detail a series of experiments that convinced him that there are two selves—the *experiencing self* and the *remembering self*—whose preferences are in conflict.

The experiencing self is the one being measured by the *hedonimeter*, which the nineteenth-century British economist Francis Edgeworth imagined to be "an ideally perfect instrument, a psychophysical machine, continually registering the height of pleasure experienced by an individual, exactly according to the verdict of consciousness."[44] According to hedonic utilitarianism, the overall value of any experience to an individual is simply the sum of the hedonic values of each instant during the experience. This notion applies equally well to eating an ice cream or living an entire life.

The remembering self, on the other hand, is the one who is "in charge" when there is any decision to be made. This self chooses new experiences based on *memories* of previous experiences and their desirability. Kahneman's experiments suggest that the remembering self has very different ideas from the experiencing self.

The simplest experiment to understand involves plunging a subject's hand into cold water. There are two different regimes: in the first, the immersion is for 60 seconds in water at 14 degrees Celsius; in the second, the immersion is for 60 seconds in water at 14 degrees followed by 30 seconds at 15 degrees. (These temperatures are similar to ocean temperatures in Northern California—cold enough that almost everyone wears a wetsuit in the water.) All subjects report the experience as unpleasant. After experiencing both regimes (in either order, with a 7-minute gap in between), the subject is asked to choose which one they would like to repeat. The great majority of subjects prefer to repeat the 60 + 30 rather than just the 60-second immersion.

Kahneman posits that, from the point of view of the experiencing self, 60 + 30 has to be *strictly worse* than 60, because it includes 60 *and another unpleasant experience.* Yet the remembering self chooses 60 + 30. Why?

Kahneman's explanation is that the remembering self looks back with rather weirdly tinted spectacles, paying attention mainly to the "peak" value (the highest or lowest hedonic value) and the "end" value (the hedonic value at the end of the experience). The durations of different parts of the experience are mostly neglected. The peak discomfort levels for 60 and 60 + 30 are the same, but the end levels are different: in the 60 + 30 case, the water is one degree warmer. If the remembering self evaluates experiences by the peak and end values, rather than by summing up hedonic values over time, then 60 + 30 is better, and this is what is found. The peak-end model seems to explain many other equally weird findings in the literature on preferences.

Kahneman seems (perhaps appropriately) to be of two minds about his findings. He asserts that the remembering self "simply made a mistake" and chose the wrong experience because its memory is faulty and incomplete; he regards this as "bad news for believers in the rationality of choice." On the other hand, he writes, "A theory of well-being that ignores what people want cannot be sustained." Suppose, for example, that Harriet has tried Pepsi and Coke and now strongly prefers Pepsi; it would be absurd to force her to drink Coke based on adding up secret hedonimeter readings taken during each trial.

The fact is that no law *requires* our preferences between experiences to be defined by the sum of hedonic values over instants of time.

It is true that standard mathematical models focus on maximizing a sum of rewards,[45] but the original motivation for this was mathematical convenience. Justifications came later in the form of technical assumptions under which it is rational to decide based on adding up rewards,[46] but those technical assumptions need not hold in reality. Suppose, for example, that Harriet is choosing between two sequences of hedonic values: [10,10,10,10,10] and [0,0,40,0,0]. It's entirely possible that she just prefers the second sequence; no mathematical law can force her to make choices based on the sum rather than, say, the maximum.

Kahneman acknowledges that the situation is complicated still further by the crucial role of anticipation and memory in well-being. The memory of a single, delightful experience—one's wedding day, the birth of a child, an afternoon spent picking blackberries and making jam—can carry one through years of drudgery and disappointment. Perhaps the remembering self is evaluating not just the experience per se but its total effect on life's future value through its effect on future memories. And presumably it's the remembering self and not the experiencing self that is the best judge of what will be remembered.

Time and change

It goes almost without saying that sensible people in the twenty-first century would not want to emulate the preferences of, say, Roman society in the second century, replete with gladiatorial slaughter for public entertainment, an economy based on slavery, and brutal massacres of defeated peoples. (We need not dwell on the obvious parallels to these characteristics in modern society.) Standards of morality clearly evolve over time as our civilization progresses—or drifts, if you prefer. This suggests, in turn, that future generations might find utterly repulsive our current attitudes to, say, the well-being of animals. For this reason, it is important that machines charged with implementing human preferences be able to respond to changes in those preferences over time rather than fixing them in stone. The three principles from Chapter 7 accommodate such changes in a natural way, because they require machines to learn and implement the current preferences of current humans—lots of them, all different—rather than a single idealized set of preferences or the preferences of machine designers who may be long dead.[47]

The possibility of changes in the typical preferences of human populations over historical time naturally focuses attention on the question of how each individual's preferences are formed and the plasticity of adult preferences. Our preferences are certainly influenced by our biology: we usually avoid pain, hunger, and thirst, for example. Our biology has remained fairly constant, however, so the remaining preferences must arise from cultural and family influences. Quite possibly, children are constantly running some form of inverse reinforcement learning to identify the preferences of parents and peers in order to explain their behavior; children then adopt these preferences as their own. Even as adults, our preferences evolve through the influence of the media, government, friends, employers, and our own direct experiences. It may be the case, for example, that many supporters of the Third Reich did not start out as genocidal sadists thirsting for racial purity.

Preference change presents a challenge for theories of rationality at both the individual and societal level. For example, Harsanyi's principle of preference autonomy seems to say that everyone is entitled to whatever preferences they have and no one else should touch them. Far from being untouchable, however, preferences are touched and modified all the time, by every experience a person has. Machines cannot help but modify human preferences, because machines modify human experiences.

It's important, although sometimes difficult, to separate preference change from preference update, which occurs when an initially uncertain Harriet learns more about her own preferences through experience. Preference update can fill in gaps in self-knowledge and perhaps add definiteness to preferences that were previously weakly held and provisional. Preference change, on the other hand, is not a process that results from additional evidence about what one's preferences actually are. In the extreme case, you can imagine it as resulting from drug administration or even brain surgery—it occurs from processes we may not understand or agree with.

Preference change is problematic for at least two reasons. The first reason is that it's not clear *which* preferences should hold sway when making a decision: the preferences that Harriet has at the time of the decision or the preferences that she will have during and after the events that result from her decision. In bioethics, for example, this is a very real dilemma because people's preferences about medical interventions and end-of-life care do change, often dramatically, after they become seriously ill.[48] Assuming these changes do not result from diminished intellectual capacity, whose preferences should be respected?[49]

The second reason that preference change is problematic is that there seems to be no obvious rational basis for changing (as opposed to updating) one's preferences. If Harriet prefers A to B, but could choose to undergo an experience that she knows will result in her preferring B to A, why would she ever do that? The outcome would be that she would then choose B, which she currently does not want.

The issue of preference change appears in dramatic form in the legend of Ulysses and the Sirens. The Sirens were mythical beings whose singing lured sailors to their doom on the rocks of certain islands in the Mediterranean. Ulysses, wishing to hear the song, ordered his sailors to plug their ears with wax and to bind him to the mast; under no circumstances were they to obey his subsequent entreaties to release him. Obviously, he wanted the sailors to respect the preferences he had initially, not the preferences he would have after the Sirens bewitch him. This legend became the title of a book by the Norwegian philosopher Jon Elster,[50] dealing with weakness of will and other challenges to the theoretical idea of rationality.

Why might an intelligent machine deliberately set out to modify the preferences of humans? The answer is quite simple: to make the preferences easier to satisfy. We saw this in Chapter 1 with the case of social-media click-through optimization. One response might be to say that machines must treat human preferences as sacrosanct: nothing can be allowed to change the human's preferences. Unfortunately, this is completely impossible. The very existence of a useful robot aide is likely to have an effect on human preferences.

One possible solution is for machines to learn about human *meta-preferences*—that is, preferences about what kinds of preference change processes might be acceptable or unacceptable. Notice the use of "preference change processes" rather than "preference changes" here. That's because wanting one's preferences to change in a specific direction often amounts to having that preference already; what's really wanted in such a case is the ability to be better at *implementing* the preference. For example, if Harriet says, "I want my preferences to change so that I don't want cake as much as I do now," then she already has a preference for a future with less cake consumption; what she really wants is to alter her cognitive architecture so that her behavior more closely reflects that preference.

By "preferences about what kinds of preference change processes might be acceptable or unacceptable," I mean, for example, a view that one may end up with "better" preferences by traveling the world and experiencing a wide variety of cultures, or by participating in a vibrant intellectual community that thoroughly explores a wide range of moral traditions, or by setting aside some hermit time for introspection and hard thinking about life and its meaning. I'll call these processes *preference-neutral*, in the sense that one does not anticipate that the process will change one's preferences in any particular direction, while recognizing that some may strongly disagree with that characterization.

Of course, not all preference-neutral processes are desirable— for example, few people expect to develop "better" preferences by whacking themselves on the head. Subjecting oneself to an acceptable process of preference change is analogous to running an experiment to find out something about how the world works: you never know in advance how the experiment will turn out, but you expect, nonetheless, to be better off in your new mental state.

The idea that there are acceptable routes to preference modification seems related to the idea that there are acceptable methods of behavior modification whereby, for example, an employer engineers the choice situation so that people make "better" choices about saving for retirement. Often this can be done by manipulating the "non-rational" factors that influence choice, rather than by restricting choices or taxing "bad" choices. *Nudge*, a book by economist Richard Thaler and legal scholar Cass Sunstein, lays out a wide range of supposedly acceptable methods and opportunities to "influence people's behavior in order to make their lives longer, healthier, and better."

It's unclear whether behavior modification methods are really just modifying behavior. If, when the nudge is removed, the modified behavior persists—which is presumably the desired outcome of such interventions—then something has changed in the individual's cognitive architecture (the thing that turns underlying preferences into behavior) or in the individual's underlying preferences. It's quite likely to be a bit of both. What is clear, however, is that the nudge strategy is assuming that everyone shares a preference for "longer, healthier, and better" lives; each nudge is based on a particular definition of a "better" life, which seems to go against the grain of preference autonomy. It might be better, instead, to design preference-neutral assistive processes that help people bring their decisions and their cognitive architectures into better alignment with their underlying preferences. For example, it's possible to design cognitive aides that highlight the longer-term consequences of decisions and teach people to recognize the seeds of those consequences in the present.[51]

That we need a better understanding of the processes whereby human preferences are formed and shaped seems obvious, not least because such an understanding would help us design machines that avoid accidental and undesirable changes in human preferences of the kind wrought by social-media content selection algorithms. Armed with such an understanding, of course, we will be tempted to engineer changes that would result in a "better" world.

Some might argue that we should provide much greater opportunities for preference-neutral "improving" experiences such as travel, debate, and training in analytical and critical thinking. We might, for example, provide opportunities for every high-school student to live for a few months in at least two other cultures distinct from his or her own.

Almost certainly, however, we will want to go further—for example, by instituting social and educational reforms that increase the coefficient of altruism—the weight that each individual places on the welfare of others—while decreasing the coefficients of sadism, pride, and envy. Would this be a good idea? Should we recruit our machines to help in the process? It's certainly tempting. Indeed, Aristotle himself wrote, "The main concern of politics is to engender a certain character in the citizens and to make them good and disposed to perform noble actions." Let's just say that there are risks associated with intentional preference engineering on a global scale. We should proceed with extreme caution.

PROBLEM SOLVED?

I f we succeed in creating provably beneficial AI systems, we would eliminate the risk that we might lose control over superintelligent machines. Humanity could proceed with their development and reap the almost unimaginable benefits that would flow from the ability to wield far greater intelligence in advancing our civilization. We would be released from millennia of servitude as agricultural, industrial, and clerical robots and we would be free to make the best of life's potential. From the vantage point of this golden age, we would look back on our lives in the present time much as Thomas Hobbes imagined life without government: solitary, poor, nasty, brutish, and short.

Or perhaps not. Bondian villains may circumvent our safeguards and unleash uncontrollable superintelligences against which humanity has no defense. And if we survive that, we may find ourselves gradually enfeebled as we entrust more and more of our knowledge and skills to machines. The machines may advise us not to do this, understanding the long-term value of human autonomy, but we may overrule them.

Beneficial Machines

The standard model underlying a good deal of twentieth-century technology relies on machinery that optimizes a fixed, exogenously supplied objective. As we have seen, this model is fundamentally flawed. It works only if the objective is guaranteed to be complete and correct, or if the machinery can easily be reset. Neither condition will hold as AI becomes increasingly powerful.

If the exogenously supplied objective can be wrong, then it makes no sense for the machine to act as if it is always correct. Hence my proposal for beneficial machines: machines whose actions can be expected to achieve *our* objectives. Because these objectives are in us, and not in them, the machines will need to learn more about what we really want from observations of the choices we make and how we make them. Machines designed in this way will defer to humans: they will ask permission; they will act cautiously when guidance is unclear; and they will allow themselves to be switched off.

While these initial results are for a simplified and idealized setting, I believe they will survive the transition to more realistic settings. Already, my colleagues have successfully applied the same approach to practical problems such as self-driving cars interacting with human drivers.[1] For example, self-driving cars are notoriously bad at handling four-way stop signs when it's not clear who has the right of way. By formulating this as an assistance game, however, the car comes up with a novel solution: it actually backs up a little bit to show that it's definitely not planning to go first. The human understands this signal and goes ahead, confident that there will be no collision. Obviously, we human experts could have thought of this solution and programmed it into the vehicle, but that's not what happened; this is a form of communication that the vehicle invented entirely by itself.

As we gain more experience in other settings, I expect that we will be surprised by the range and fluency of machine behaviors as they interact with humans. We are so used to the stupidity of machines that execute inflexible, preprogrammed behaviors or pursue definite but incorrect objectives that we may be shocked by how sensible they become. The technology of provably beneficial machines is the core of a new approach to AI and the basis for a new relationship between humans and machines.

It seems possible, also, to apply similar ideas to the redesign of other "machines" that ought to be serving humans, beginning with ordinary software systems. We are taught to build software by composing subroutines, each of which has a well-defined *specification* that says what the output should be for any given input—just like the square-root button on a calculator. This specification is the direct analog of the objective given to an AI system. The subroutine is not supposed to terminate and return control to the higher layers of the software system until it has produced an output that meets the specification. (This should remind you of the AI system that persists in its single-minded pursuit of its given objective.) A better approach would be to allow for uncertainty in the specification. For example, a subroutine that carries out some fearsomely complicated mathematical computation is typically given an error bound that defines the required precision for the answer and has to return a solution that is correct within that error bound. Sometimes, this may require weeks of computation. Instead, it might be better to be less precise about the allowed error, so that the subroutine could come back after twenty seconds and say, "I've found a solution that's *this* good. Is that OK or do you want me to continue?" In some cases, the question may percolate all the way to the top level of the software system, so that the human user can provide further guidance to the system. The human's answers would then help in refining the specifications at all levels.

The same kind of thinking can be applied to entities such as governments and corporations. The obvious failings of government include paying too much attention to the preferences (financial as well as political) of those in government and too little attention to the preferences of the governed. Elections are supposed to communicate preferences to the government, but they seem to have a remarkably small bandwidth (on the order of one byte of information every few years) for such a complex task. In far too many countries, government is simply a means for one group of people to impose its will on others. Corporations go to greater lengths to learn the preferences of customers, whether through market research or direct feedback in the form of purchase decisions. On the other hand, the molding of human preferences through advertising, cultural influences, and even chemical addiction is an accepted way of doing business.

Governance of AI

AI has the power to reshape the world, and the process of reshaping will have to be managed and guided in some way. If the sheer number of initiatives to develop effective governance of AI is any guide, then we are in excellent shape. Everyone and their uncle is setting up a Board or a Council or an International Panel. The World Economic Forum has identified nearly three hundred separate efforts to develop ethical principles for AI. My email inbox can be summarized as one long invitation to the Global World Summit Conference Forum on the Future of International Governance of the Social and Ethical Impacts of Emerging Artificial Intelligence Technologies.

This is all very different from what happened with nuclear technology. After World War II, the United States held all the nuclear cards. In 1953, US president Dwight Eisenhower proposed to the UN an international body to regulate nuclear technology. In 1957, the International Atomic Energy Agency started work; it is the sole global overseer for the safe and beneficial development of nuclear energy.

In contrast, many hands hold AI cards. To be sure, the United

States, China, and the EU fund a lot of AI research, but almost all of it occurs outside secure national laboratories. AI researchers in universities are part of a broad, cooperative international community, glued together by shared interests, conferences, cooperative agreements, and professional societies such as AAAI (the Association for the Advancement of Artificial Intelligence) and IEEE (the Institute of Electrical and Electronics Engineers, which includes tens of thousands of AI researchers and practitioners). Probably the majority of investment in AI research and development is now occurring within corporations, large and small; the leading players as of 2019 are Google (including DeepMind), Facebook, Amazon, Microsoft, and IBM in the United States and Tencent, Baidu, and, to some extent, Alibaba in China—all among the largest corporations in the world.[2] All but Tencent and Alibaba are members of the Partnership on AI, an industry consortium that includes among its tenets a promise of cooperation on AI safety. Finally, although the vast majority of humans possess little in the way of AI expertise, there is at least a superficial willingness among other players to take the interests of humanity into account.

These, then, are the players who hold the majority of the cards. Their interests are not in perfect alignment but all share a desire to maintain control over AI systems as they become more powerful. (Other goals, such as avoiding mass unemployment, are shared by governments and university researchers, but not necessarily by corporations that expect to profit in the short term from the widest possible deployment of AI.) To cement this shared interest and achieve coordinated action, there are organizations with *convening power*, which means, roughly, that if the organization sets up a meeting, people accept the invitation to participate. In addition to the professional societies, which can bring AI researchers together, and the Partnership on AI, which combines corporations and nonprofit institutes, the canonical conveners are the UN (for governments and researchers) and the World Economic Forum (for governments and corporations). In addition, the G7 has proposed an International Panel on Artificial

Intelligence, hoping that it will grow into something like the UN's Intergovernmental Panel on Climate Change. Important-sounding reports are multiplying like rabbits.

With all this activity, is there any prospect of actual progress on governance occurring? Perhaps surprisingly, the answer is yes, at least around the edges. Many governments around the world are equipping themselves with advisory bodies to help with the process of developing regulations; perhaps the most prominent example is the EU's High-Level Expert Group on Artificial Intelligence. Agreements, rules, and standards are beginning to emerge for issues such as user privacy, data exchange, and avoiding racial bias. Governments and corporations are working hard to sort out the rules for self-driving cars—rules that will inevitably have cross-border elements. There is a consensus that AI decisions must be explainable if AI systems are to be trusted, and that consensus is already partially implemented in the EU's GDPR legislation. In California, a new law forbids AI systems to impersonate humans in certain circumstances. These last two items— explainability and impersonation—certainly have some bearing on issues of AI safety and control.

At present, there are no implementable recommendations that can be made to governments or other organizations considering the issue of maintaining control over AI systems. A regulation such as "AI systems must be safe and controllable" would carry no weight, because these terms do not yet have precise meanings and because there is no widely known engineering methodology for ensuring safety and controllability. But let's be optimistic and imagine that, a few years down the line, the validity of the "provably beneficial" approach to AI has been established through both mathematical analysis and practical realization in the form of useful applications. We might, for example, have personal digital assistants that we can trust to use our credit cards, screen our calls and emails, and manage our finances because they have adapted to our individual preferences and know when it's OK to go ahead and when it's better to ask for guidance. Our self-driving cars may have learned good manners for interacting with one another and with human drivers, and our domestic robots should be interacting smoothly with even the most recalcitrant toddler. With luck, no cats will have been roasted for dinner and no whale meat will have been served to members of the Green Party.

At that point, it might be feasible to specify software design templates to which various kinds of applications must conform in order to be sold or connected to the Internet, just as applications have to pass a number of software tests before they can be sold on Apple's App Store or Google Play. Software vendors could propose additional templates, as long as they come with proofs that the templates satisfy the (by then well-defined) requirements of safety and controllability. There would be mechanisms for reporting problems and for updating software systems that produce undesirable behavior. It would make sense also to create professional codes of conduct around the idea of provably safe AI programs and to integrate the corresponding theorems and methods into the curriculum for aspiring AI and machine learning practitioners.

To a seasoned observer of Silicon Valley, this may sound rather naïve. Regulation of any kind is strenuously opposed in the Valley. Whereas we are accustomed to the idea that pharmaceutical companies have to show safety and (beneficial) efficacy through clinical trials before they can release a product to the general public, the software industry operates by a different set of rules—namely, the empty set. A "bunch of dudes chugging Red Bull"[3] at a software company can unleash a product or an upgrade that affects literally billions of people with no third-party oversight whatsoever.

Inevitably, however, the tech industry is going to have to acknowledge that its products matter; and, if they matter, then it matters that the products not have harmful effects. This means that there will be rules governing the nature of interactions with humans, prohibiting designs that, say, consistently manipulate preferences or produce addictive behavior. I have no doubt that the transition from an unregulated to a regulated world will be a painful one. Let's hope it doesn't require a Chernobyl-sized disaster (or worse) to overcome the industry's resistance.

Misuse

Regulation might be painful for the software industry, but it would be intolerable for Dr. Evil, plotting world domination in his secret underground bunker. There is no doubt that criminal elements, terrorists, and rogue nations would have an incentive to circumvent any constraints on the design of intelligent machines so that they could be used to control weapons or to devise and carry out criminal activities. The danger is not so much that the evil schemes would succeed; it is that they would fail by losing control over poorly designed intelligent systems—particularly ones imbued with evil objectives and granted access to weapons.

This is not a reason to avoid regulation—after all, we have laws against murder even though they are often circumvented. It does, however, create a very serious policing problem. Already, we are losing the battle against malware and cybercrime. (A recent report estimates over two billion victims and an annual cost of around $600 billion.[4]) Malware in the form of highly intelligent programs would be much harder to defeat.

Some, including Nick Bostrom, have proposed that we use our own, beneficial superintelligent AI systems to detect and destroy any malicious or otherwise misbehaving AI systems. Certainly, we should use the tools at our disposal, while minimizing the impact on personal freedom, but the image of humans huddling in bunkers, defenseless against the titanic forces unleashed by battling superintelligences, is hardly reassuring even if some of them are on our side. It would be far better to find ways to nip the malicious AI in the bud.

A good first step would be a successful, coordinated, international campaign against cybercrime, including expansion of the Budapest Convention on Cybercrime. This would form an organizational template for possible future efforts to prevent the emergence of uncontrolled AI programs. At the same time, it would engender a broad cultural understanding that creating such programs, either deliberately or inadvertently, is in the long run a suicidal act comparable to creating pandemic organisms.

Enfeeblement and Human Autonomy

E. M. Forster's most famous novels, including *Howards End* and *A Passage to India*, examined British society and its class system in the early part of the twentieth century. In 1909, he wrote one notable science-fiction story: "The Machine Stops." The story is remarkable for its prescience, including depictions of (what we would now call) the Internet, videoconferencing, iPads, massive open online courses (MOOCs), widespread obesity, and avoidance of face-to-face contact. The Machine of the title is an all-encompassing intelligent infrastructure that meets all human needs. Humans become increasingly dependent on it, but they understand less and less about how it works. Engineering knowledge gives way to ritualized incantations that eventually fail to stem the gradual deterioration of the Machine's workings. Kuno, the main character, sees what is unfolding but is powerless to stop it:

> Cannot you see . . . that it is we that are dying, and that down here the only thing that really lives is the Machine? We created the Machine to do our will, but we cannot make it do our will now. It has robbed us of the sense of space and of the sense of touch, it has blurred every human relation, it has paralysed our bodies and our wills. . . . We only exist as the blood corpuscles that course through its arteries, and if it could work without us, it would let us die. Oh,

I have no remedy—or, at least, only one—to tell men again and again that I have seen the hills of Wessex as Aelfrid saw them when he overthrew the Danes.

More than one hundred billion people have lived on Earth. They (we) have spent on the order of one trillion person-years learning and teaching, in order that our civilization may continue. Up to now, its only possibility for continuation has been through re-creation in the minds of new generations. (Paper is fine as a method of transmission, but paper does nothing until the knowledge recorded thereon reaches the next person's mind.) That is now changing: increasingly, it is possible to place our knowledge into machines that, by themselves, can run our civilization for us.

Once the practical incentive to pass our civilization on to the next generation disappears, it will be very hard to reverse the process. One trillion years of cumulative learning would, in a real sense, be lost. We would become passengers in a cruise ship run by machines, on a cruise that goes on forever—exactly as envisaged in the film *WALL-E.*

A good consequentialist would say, "Obviously this is an undesirable consequence of the overuse of automation! Suitably designed machines would never do this!" True, but think what this means. Machines may well understand that human autonomy and competence are important aspects of how we prefer to conduct our lives. They may well insist that humans retain control and responsibility for their own well-being—in other words, machines will say no. But we myopic, lazy humans may disagree. There is a tragedy of the commons at work here: for any individual human, it may seem pointless to engage in years of arduous learning to acquire knowledge and skills that machines already have; but if everyone thinks that way, the human race will, collectively, lose its autonomy.

The solution to this problem seems to be cultural, not technical. We will need a cultural movement to reshape our ideals and preferences towards autonomy, agency, and ability and away from self-indulgence and dependency—if you like, a modern, cultural version of ancient Sparta's military ethos. This would mean human preference engineering on a global scale along with radical changes in how our society works. To avoid making a bad situation worse, we might need the help of superintelligent machines, both in shaping the solution and in the actual process of achieving a balance for each individual.

Any parent of a small child is familiar with this process. Once the child is beyond the helpless stage, parenting requires an ever-evolving balance between doing everything for the child and leaving the child entirely to his or her own devices. At a certain stage, the child comes to understand that the parent is perfectly capable of tying the child's shoelaces but is choosing not to. Is that the future for the human race—to be treated like a child, forever, by far superior machines? I suspect not. For one thing, children cannot switch their parents off. (Thank goodness!) Nor will we be pets or zoo animals. There is really no analog in our present world to the relationship we will have with beneficial intelligent machines in the future. It remains to be seen how the endgame turns out.

Appendix A

SEARCHING FOR SOLUTIONS

Choosing an action by looking ahead and considering the outcomes of different possible action sequences is a fundamental capability for intelligent systems. It's something your cell phone does whenever you ask it for directions. Figure 14 shows a typical example: getting from the current location, Pier 19, to the goal, Coit Tower. The algorithm needs to know what actions are available to it; typically, for map navigation, each action traverses a road segment connecting two adjacent intersections. In the example, from Pier 19 there is just one action: turn right and drive along the Embarcadero to the next intersection. Then there is a choice: continue on or take a sharp left onto Battery Street. The algorithm systematically explores all these possibilities until it eventually finds a route. Typically we add a little bit of commonsense guidance, such as a preference for exploring streets that head towards the goal rather than away from it. With this guidance and a few other tricks, the algorithm can find optimal solutions very quickly—usually in a few milliseconds, even for a cross-country trip.

Searching for routes on maps is a natural and familiar example, but it may be a bit misleading because the number of distinct locations is so small. In the United States, for example, there are only about ten

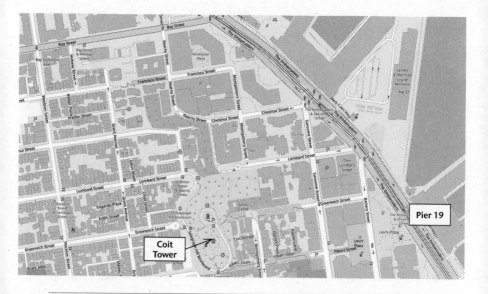

FIGURE 14: A map of part of San Francisco, showing the initial location at Pier 19 and the destination at Coit Tower.

million intersections. That may seem like a large number, but it is tiny compared to the number of distinct states in the 15-puzzle. The 15-puzzle is a toy with a four-by-four grid containing fifteen numbered tiles and a blank space. The goal is to move the tiles around to achieve a goal configuration, such as having all the tiles in numerical order. The 15-puzzle has about ten trillion states (a million times bigger than the United States!); the 24-puzzle has about eight trillion trillion states. This is an example of what mathematicians call *combinatorial complexity*—the rapid explosion in the number of combinations as the number of "moving parts" of a problem increases. Returning to the map of the United States: if a trucking company wants to optimize the movements of its one hundred trucks across the United States, the number of possible states to consider would be ten million to the power of one hundred (i.e., 10^{700}).

Giving up on rational decisions

Many games have this property of combinatorial complexity, including chess, checkers, backgammon, and Go. Because the rules of Go are simple and elegant (figure 15), I'll use it as a running example. The objective is clear enough: win the game by surrounding more territory than your opponent. The possible actions are clear too: put a stone in an empty location. Just as with navigation on a map, the obvious way to decide what to do is to imagine different futures that result from different sequences of actions and choose the best one. You ask, "If I do this, what might my opponent do? And what do I do then?" This idea is illustrated in figure 16 for 3×3 Go. Even for 3×3 Go, I can show only a small part of the tree of possible futures, but I hope the idea is clear enough. Indeed, this way of making decisions seems to be just straightforward common sense.

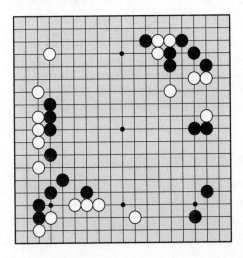

FIGURE 15: A Go board, partway through Game 5 of the 2002 LG Cup final between Lee Sedol (black) and Choe Myeong-hun (white). Black and White take turns placing a single stone on any unoccupied location on the board. Here, it is Black's turn to move and there are 343 possible moves. Each side attempts to surround as much territory as possible. For example, White has good chances to win territory at the left-hand edge and on the left side of the bottom edge, while Black may win territory in the top-right and bottom-right corners. A key concept in Go is that of a *group*— that is, a set of stones of the same color that are connected to one another by vertical or horizontal adjacency. A group remains alive as long as there is at least one empty space next to it; if it is completely surrounded, with no empty spaces, it dies and is removed from the board.

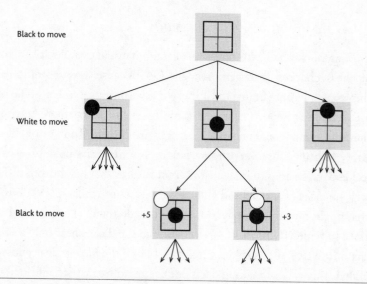

Black to move

White to move

Black to move   +5   +3

FIGURE 16: Part of the game tree for 3×3 Go. Starting from the empty initial state, sometimes called the *root* of the tree, Black can choose one of three possible distinct moves. (The others are symmetric with these.) It would then be White's turn to move. If Black chooses to play in the center, White has two distinct moves—corner or side—then Black would get to play again. By imagining these possible futures, Black can choose which move to play in the initial state. If Black is unable to follow every possible line of play to the end of the game, then an evaluation function can be used to estimate how good the positions are at the leaves of the tree. Here, the evaluation function assigns +5 and +3 to two of the leaves.

The problem is that Go has more than 10^{170} possible positions for the full 19×19 board. Whereas finding a guaranteed shortest route on a map is relatively easy, finding a guaranteed win in Go is utterly infeasible. Even if the algorithm ponders for the next billion years, it can explore only a tiny fraction of the whole tree of possibilities. This leads to two questions. First, which part of the tree should the program explore? And second, which move should the program make, given the partial tree that it has explored?

To answer the second question first: the basic idea used by almost all lookahead programs is to assign an *estimated value* to the "leaves" of the tree—those states furthest in the future—and then "work back" to find out how good the choices are at the root.[1] For example, looking at the two positions at the bottom of figure 16, one might guess a value of +5 (from Black's viewpoint) for the position on the left and +3 for the position on the right, because White's stone in the corner is much more vulnerable than the one on the side. If these values are right, then Black can expect that White will play on the side, leading to the right-hand position; hence, it seems reasonable to assign a value of +3 to Black's initial move in the center. With slight variations, this is the scheme used by Arthur Samuel's checker-playing program to beat its creator in 1955,[2] by Deep Blue to beat the then world chess champion, Garry Kasparov, in 1997, and by AlphaGo to beat former world Go champion Lee Sedol in 2016. For Deep Blue, humans wrote the piece of the program that evaluates positions at the leaves of the tree, based largely on their knowledge of chess. For Samuel's program and for AlphaGo, the programs learned it from thousands or millions of practice games.

The first question—which part of the tree should the program explore?—is an example of one of the most important questions in AI: *What computations should an agent do?* For game-playing programs, it is vitally important because they have only a small, fixed allocation of time, and using it on pointless computations is a sure way to lose. For humans and other agents operating in the real world, it is even more important because the real world is so much more complex: unless chosen well, no amount of computation is going to make the smallest dent in the problem of deciding what to do. If you are driving and a moose walks into the middle of the road, it's no use thinking about whether to trade euros for pounds or whether Black should make its first move in the center of the Go board.

The ability of humans to manage their computational activity so that reasonable decisions get made reasonably quickly is at least as remarkable as their ability to perceive and to reason correctly. And it seems to be something we acquire naturally and effortlessly: when my father taught me to play chess, he taught me the rules, but he did not also teach me such-and-such clever algorithm for choosing which parts of the game tree to explore and which parts to ignore.

How does this happen? On what basis can we direct our thoughts? The answer is that a computation has value to the extent that it can improve your decision quality. The process of choosing computations is called *metareasoning*, which means reasoning about reasoning. Just as actions can be chosen rationally, on the basis of expected value, so can computations. This is called *rational metareasoning*.[3] The basic idea is very simple:

> Do the computations that will give the highest expected improvement in decision quality, and stop when the cost (in terms of time) exceeds the expected improvement.

That's it. No fancy algorithm needed! This simple principle generates effective computational behavior in a wide range of problems, including chess and Go. It seems likely that our brains implement something similar, which explains why we don't need to learn new, game-specific algorithms for thinking with each new game we learn to play.

Exploring a tree of possibilities that stretches forward into the future from the current state is not the only way to reach decisions, of course. Often, it makes more sense to work backwards from the goal. For example, the presence of the moose in the road suggests the goal of *avoid hitting the moose*, which in turn suggests three possible actions: swerve left, swerve right, or slam on the brakes. It does not suggest the action of trading euros for pounds or putting a black stone in the center. Thus, goals have a wonderful focusing effect on one's thinking. No current game-playing programs take advantage of this idea; in fact, they typically consider all possible legal actions. This is one of the (many) reasons why I am not worried about AlphaZero taking over the world.

Looking further ahead

Let's suppose you have decided to make a specific move on the Go board. Great! Now you have to actually do it. In the real world, this involves reaching into the bowl of unplayed stones to pick up a stone, moving your hand above the intended location, and placing the stone neatly on the spot, either quietly or emphatically according to Go etiquette.

Each of these stages, in turn, consists of a complex dance of perception and motor control commands involving the muscles and nerves of the hand, arm, shoulder, and eyes. And while reaching for a stone, you're making sure the rest of your body doesn't topple over thanks to the shift in your center of gravity. The fact that you may not be consciously aware of selecting these actions does not mean that they aren't being selected by your brain. For example, there may be many stones in the bowl, but your "hand"—really, your brain processing sensory information—still has to choose one of them to pick up.

Almost everything we do is like this. While driving, we might choose to *change lanes to the left*; but this action involves looking in the mirror and over your shoulder, perhaps adjusting speed, and moving the steering wheel while monitoring progress until the maneuver is complete. In conversation, a routine response such as "OK, let me check my calendar and get back to you" involves articulating fourteen syllables, each of which requires hundreds of precisely coordinated motor control commands to the muscles of the tongue, lips, jaw, throat, and breathing apparatus. For your native language, this process is automatic; it closely resembles the idea of running a subroutine in a computer program (see page 34). The fact that complex action sequences can become routine and automatic, thereby functioning as single actions in still more complex processes, is absolutely fundamental to human cognition. Saying words in a less familiar language—perhaps asking directions to Szczebrzeszyn in Poland—is a useful reminder that there was a time in your life when reading and speaking words were difficult tasks requiring mental effort and lots of practice.

So, the real problem that your brain faces is not choosing a move on the Go board but sending motor control commands to your muscles. If we shift our attention from the level of Go moves to the level of motor control commands, the problem looks very different. Very roughly, your brain can send out commands about every one hundred milliseconds. We have about six hundred muscles, so that's a theoretical maximum of about six thousand actuations per second, twenty million per hour, two hundred billion per year, twenty trillion per lifetime. Use them wisely!

Now, suppose we tried to apply an AlphaZero-like algorithm to solve the decision problem at this level. In Go, AlphaZero looks ahead perhaps fifty steps. But fifty steps of motor control commands get you only a few seconds into the future! Not enough for the twenty million motor control commands in an hour-long game of Go, and certainly not enough for the trillion (1,000,000,000,000) steps involved in doing a PhD. So, even though AlphaZero looks further ahead in Go than any human can, that ability doesn't seem to help in the real world. It's the wrong kind of lookahead.

I'm not saying, of course, that doing a PhD actually requires planning out a trillion muscle actuations in advance. Only quite abstract plans are made initially—perhaps choosing Berkeley or some other place, choosing a PhD supervisor or research topic, applying for funding, getting a student visa, traveling to the chosen city, doing some research, and so on. To make your choices, you do just enough thinking, about just the right things, so that the decision becomes clear. If the feasibility of some abstract step such as getting the visa is unclear, you do some more thinking and perhaps information gathering, which means making the plan more concrete in certain aspects: maybe choosing a visa type for which you are eligible, collecting the necessary documents, and submitting the application. Figure 17 shows the abstract plan and the refinement of the GetVisa step into a three-step subplan. When the time comes to begin carrying out the plan, its initial steps have to be refined all the way down to the primitive level so that your body can execute them.

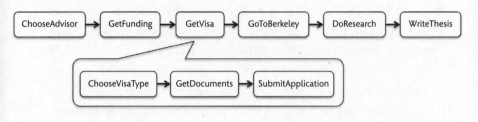

FIGURE 17: An abstract plan for an overseas student who has chosen to get a PhD at Berkeley. The GetVisa step, whose feasibility is uncertain, has been expanded out into an abstract plan of its own.

AlphaGo simply cannot do this kind of thinking: the only actions it ever considers are primitive actions occurring in a sequence from the initial state. It has no notion of *abstract plan*. Trying to apply AlphaGo in the real world is like trying to write a novel by wondering whether the first letter should be A, B, C, and so on.

In 1962, Herbert Simon emphasized the importance of hierarchical organization in a famous paper, "The Architecture of Complexity."[4] AI researchers since the early 1970s have developed a variety of methods that construct and refine hierarchically organized plans.[5] Some of the resulting systems are able to construct plans with tens of millions of steps—for example, to organize manufacturing activities in a large factory.

We now have a pretty good theoretical understanding of the meaning of abstract actions—that is, of how to define the effects they have on the world.[6] Consider, for example, the abstract action GoToBerkeley in figure 17. It can be implemented in many different ways, each of which produces different effects on the world: you could sail there, stow away on a ship, fly to Canada and walk across the border, hire a private jet, and so on. But you need not consider any of these choices for now. As long as you are sure there is a way to do it that doesn't consume so much time and money or incur so much risk as to imperil the rest of the plan, you can just put the abstract step GoToBerkeley into the plan and rest assured that the plan will work. In this way, we can build high-level plans that will eventually turn into billions or trillions of primitive steps without ever worrying about what those steps are until it's time to actually do them.

Of course, none of this is possible without the hierarchy. Without high-level actions such as getting a visa and writing a thesis, we cannot make an abstract plan to get a PhD; without still-higher-level actions such as getting a PhD and starting a company, we cannot plan to get a PhD and then start a company. In the real world, we would be lost without a vast library of actions at dozens of levels of abstraction. (In the game of Go, there is no obvious hierarchy of actions, so most of us *are* lost.) At present, however, all existing methods for hierarchical planning rely on a human-generated hierarchy of abstract and concrete actions; we do not yet understand how such hierarchies can be learned from experience.

KNOWLEDGE AND LOGIC

L ogic is the study of reasoning with definite knowledge. It is fully general with regard to subject matter—that is, the knowledge can be about anything at all. Logic is therefore an indispensable part of our understanding of general purpose intelligence.

Logic's main requirement is a *formal* language with precise meanings for the sentences in the language, so that there is an unambiguous process for determining whether a sentence is true or false in a given situation. That's it. Once we have that, we can write *sound* reasoning algorithms that produce new sentences from sentences that are already known. Those new sentences are guaranteed to follow from the sentences that the system already knows, meaning that the new sentences are necessarily true in any situation where the original sentences are true. This allows a machine to answer questions, prove mathematical theorems, or construct plans that are guaranteed to succeed.

High-school algebra provides a good example (albeit one that may evoke painful memories). The formal language includes sentences such as $4x + 1 = 2y - 5$. This sentence is true in the situation where $x = 5$ and $y = 13$, and false when $x = 5$ and $y = 6$. From this sentence one can derive another sentence such as $y = 2x + 3$, and whenever the first sentence is true, the second is guaranteed to be true too.

The core idea of logic, developed independently in ancient India, China, and Greece, is that the same notions of precise meaning and sound reasoning can be applied to sentences about anything at all, not just numbers. The canonical example starts with "Socrates is a man" and "All men are mortal" and derives "Socrates is mortal."[1] This derivation is strictly formal in the sense that it does not rely on any further information about who Socrates is or what *man* and *mortal* mean. The fact that logical reasoning is strictly formal means that it is possible to write algorithms that do it.

Propositional logic

For our purposes in understanding the capabilities and prospects for AI, there are two important kinds of logic that really matter: propositional logic and first-order logic. The difference between the two is fundamental to understanding the current situation in AI and how it is likely to evolve.

Let's start with propositional logic, which is the simpler of the two. Sentences are made of just two kinds of things: symbols that stand for propositions that can be true or false, and logical *connectives* such as **and**, **or**, **not**, and **if . . . then**. (We'll see an example shortly.) These logical connectives are sometimes called *Boolean*, after George Boole, a nineteenth-century logician who reinvigorated his field with new mathematical ideas. They are just the same as the *logic gates* used in computer chips.

Practical algorithms for reasoning in propositional logic have been known since the early 1960s.[2,3] Although the general reasoning task may require exponential time in the worst case,[4] modern propositional reasoning algorithms handle problems with millions of proposition symbols and tens of millions of sentences. They are a core tool for constructing guaranteed logistical plans, verifying chip designs before they are manufactured, and checking the correctness of software applications and security protocols before they are deployed. The amazing thing is that a single algorithm—a reasoning algorithm for propositional logic—solves *all* these tasks once they have been formulated as reasoning tasks. Clearly, this is a step towards the goal of generality in intelligent systems.

Unfortunately, it's not a very big step because the language of propositional logic is not very expressive. Let's see what this means in practice when we try to express the basic rule for legal moves in Go: "The player whose turn it is to move can play a stone on any unoccupied intersection."[5] The first step is to decide what the proposition symbols are going to be for talking about Go moves and Go board positions. The fundamental proposition that matters is whether a stone of a particular color is on a particular location at a particular time. So, we'll need symbols such as *White_Stone_On_5_5_At_Move_38* and *Black_Stone_On_5_5_At_Move_38*. (Remember that, as with *man, mortal,* and *Socrates,* the reasoning algorithm doesn't need to know what the symbols mean.) Then the logical condition for White to be able to play at the 5,5 intersection at move 38 would be

(not *White_Stone_On_5_5_At_Move_38*) **and**
(not *Black_Stone_On_5_5_At_Move_38*)

In other words: there's no white stone and there's no black stone. That seems simple enough. Unfortunately, in propositional logic it would have to be written out separately for each location and for each move in the game. Because there are 361 locations and around 300 moves per game, this means over 100,000 copies of the rule! For the rules concerning captures and repetitions, which involve multiple stones and locations, the situation is even worse, and we quickly fill up millions of pages.

The real world is, obviously, much bigger than the Go board: there are far more than 361 locations and 300 time steps, and there are many kinds of things besides stones; so, the prospect of using a propositional language for knowledge of the real world is utterly hopeless.

It's not just the ridiculous *size* of the rulebook that's a problem: it's also the ridiculous amount of *experience* a learning system would need to acquire the rules from examples. While a human needs just one or two examples to get the basic ideas of placing a stone, capturing stones, and so on, an intelligent system based on propositional logic has to be shown examples of moving and capturing separately for each location and time step. The system cannot generalize from a few examples, as a human does, because it has no way to express the general rule. This limitation applies not just to systems based on propositional logic but also to any system with comparable expressive power. That includes Bayesian networks, which are probabilistic cousins of propositional logic, and neural networks, which are the basis for the "deep learning" approach to AI.

First-order logic

So, the next question is, can we devise a more expressive logical language? We'd like one in which it is possible to tell the rules of Go to the knowledge-based system in the following way:

> **for all** locations on the board, and **for all** time steps, here are the rules . . .

First-order logic, introduced by the German mathematician Gottlob Frege in 1879, allows one to write the rules this way.[6] The key difference between propositional and first-order logic is this: whereas propositional logic assumes the world is made of propositions that are true or false, first-order logic assumes the world is made of *objects* that can be *related* to each other in various ways. For example, there could be locations that are adjacent to each other, times that follow each other consecutively, stones that are on locations at particular times, and moves that are legal at particular times. First-order logic allows one to assert that some property is true for *all* objects in the world; so, one can write

> **for all** time steps *t*, and **for all** locations *l*, and **for all** colors *c*,
>
> if it is *c*'s turn to move at time *t* **and** *l* is unoccupied at time *t*,
>
> **then** it is legal for *c* to play a stone at location *l* at time *t*.

With some extra caveats and some additional sentences that define the board locations, the two colors, and what *unoccupied* means, we have the beginnings of the complete rules of Go. The rules take up about as much space in first-order logic as they do in English.

The development of *logic programming* in the late 1970s provided elegant and efficient technology for logical reasoning embodied in a programming language called Prolog. Computer scientists worked out how to make logical reasoning in Prolog run at millions of reasoning steps per second, making many applications of logic practical. In 1982, the Japanese government announced a huge investment in Prolog-based AI called the Fifth Generation project,[7] and the United States and UK responded with similar efforts.[8,9]

Unfortunately, the Fifth Generation project and others like it ran out of steam in the late 1980s and early 1990s, partly because of the inability of logic to handle uncertain information. They epitomized what soon became a pejorative term: *Good Old-Fashioned AI*, or GOFAI.[10] It became fashionable to dismiss logic as irrelevant to AI; indeed, many AI researchers working now in the area of deep learning don't know anything about logic. This fashion seems likely to fade: if you accept that the world has objects in it that are related to each other in various ways, then first-order logic is going to be relevant, because it provides the basic mathematics of objects and relations. This view is shared by Demis Hassabis, CEO of Google DeepMind:[11]

> You can think about deep learning as it currently is today as the equivalent in the brain to our sensory cortices: our visual cortex or auditory cortex. But, of course, true intelligence is a lot more than just that, you have to recombine it into higher-level thinking and

symbolic reasoning, a lot of the things classical AI tried to deal with in the 80s.

 . . . We would like [these systems] to build up to this symbolic level of reasoning—maths, language, and logic. So that's a big part of our work.

Thus, one of the most important lessons from the first thirty years of AI research is that a program that knows things, in any useful sense, will need a capacity for representation and reasoning that is at least comparable to that offered by first-order logic. As yet, we do not know the exact form this will take: it may be incorporated into probabilistic reasoning systems, into deep learning systems, or into some still-to-be-invented hybrid design.

UNCERTAINTY AND PROBABILITY

Whereas logic provides a general basis for reasoning with definite knowledge, probability theory encompasses reasoning with uncertain information (of which definite knowledge is a special case). Uncertainty is the normal epistemic situation of an agent in the real world. Although the basic ideas of probability were developed in the seventeenth century, only recently has it become possible to represent and reason with large probability models in a formal way.

The basics of probability

Probability theory shares with logic the idea that there are possible worlds. One usually starts out by defining what they are—for example, if I am rolling one ordinary six-sided die, there are six worlds (sometimes called *outcomes*): 1, 2, 3, 4, 5, 6. Exactly one of them will be the case, but a priori I don't know which. Probability theory assumes that it is possible to attach a probability to each world; for my die roll, I'll attach ⅙ to each world. (These probabilities happen to be equal, but it need not be that way; the only requirement is that the probabilities have to add up to 1.) Now I can ask a question such as "What's the probability I'll roll an even number?" To find this, I simply add up the probabilities for the three worlds where the number is even: $\frac{1}{6} + \frac{1}{6} + \frac{1}{6} = \frac{1}{2}$.

It's also straightforward to take new evidence into account. Suppose an oracle tells me that the roll is a prime number (that is, 2, 3, or 5). This rules out the worlds 1, 4, and 6. I simply take the probabilities associated with the remaining possible worlds and scale them up so the total remains 1. Now the probabilities of 2, 3, and 5 are each $\frac{1}{3}$, and the probability that my roll is an even number is now just $\frac{1}{3}$, since 2 is the only remaining even roll. This process of updating probabilities as new evidence arrives is an example of Bayesian updating.

So, this probability stuff seems quite simple! Even a computer can add up numbers, so what's the problem? The problem comes when there are more than a few worlds. For example, if I roll the die one hundred times, there are 6^{100} outcomes. It's infeasible to begin the process of probabilistic reasoning by attaching a number to each of these outcomes individually. A clue for dealing with this complexity comes from the fact that the die rolls are *independent* if the die is known to be fair—that is, the outcome of any single roll does not affect the probabilities for the outcomes of any other roll. Thus, independence is helpful in structuring the probabilities for complex sets of events.

Suppose I am playing Monopoly with my son George. My piece is on Just Visiting, and George owns the yellow set whose properties are sixteen, seventeen, and nineteen squares away from Just Visiting. Should he buy houses for the yellow set now, so that I have to pay him some exorbitant rent if I land on those squares, or should he wait until the next turn? That depends on the probability of landing on the yellow set in my current turn.

Here are the rules for rolling the dice in Monopoly: two dice are rolled and the piece is moved according to the total shown; if doubles are rolled, the player rolls again and moves again; if the second roll is doubles, the player rolls a third time and moves again (but if the third roll is doubles, the player goes to jail instead). So, for example, I might roll 4-4 followed by 5-4, totaling 17; or 2-2, then 2-2, then 6-2, totaling 16. As before, I simply add up the probabilities of all worlds where I land on the yellow set. Unfortunately, there are a lot of worlds. As many as six dice could be rolled altogether, so the number of worlds runs into the thousands. Furthermore, the rolls are no longer independent, because the second roll won't exist unless the first roll is doubles. On the other hand, if we fix the values of the first pair of dice, then the values of the second pair of dice are independent. Is there a way to capture this kind of dependency?

Bayesian networks

In the early 1980s, Judea Pearl proposed a formal language called *Bayesian networks* (often abbreviated to Bayes nets) that makes it possible, in many real-world situations, to represent the probabilities of a very large number of outcomes in a very concise form.[1]

Figure 18 shows a Bayesian network that describes the rolling of dice in Monopoly. The only probabilities that have to be supplied are the $1/6$ probabilities of the values 1, 2, 3, 4, 5, 6 for the individual die rolls (D_1, D_2, etc.)—that is, thirty-six numbers instead of thousands. Explaining the exact meaning of the network requires a little bit of mathematics,[2] but the basic idea is that the arrows denote *dependency* relationships—for example, the value of $Doubles_{12}$ *depends on* the values of D_1 and D_2. Similarly, the values of D_3 and D_4 (the next roll of the two dice) depend on $Doubles_{12}$ because if $Doubles_{12}$ has value *false*, then D_3 and D_4 have value 0 (that is, there is no next roll).

Just as with propositional logic, there are algorithms that can answer any question for any Bayesian network with any evidence. For example, we can ask for the probability of *LandsOnYellowSet*, which

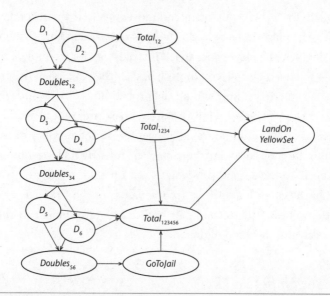

FIGURE 18: A Bayesian network that represents the rules for rolling dice in Monopoly and enables an algorithm to calculate the probability of landing on a particular set of squares (such as the yellow set) starting from some other square (such as Just Visiting). (For simplicity, the network omits the possibility of landing on a Chance or Community Chest square and being diverted to a different location.) D_1 and D_2 represent the initial roll of two dice and they are independent (no link between them). If doubles are rolled ($Doubles_{12}$), then the player rolls again, so D_3 and D_4 have non-zero values, and so on. In the situation described, the player lands on the yellow set if any of the three totals is 16, 17, or 19.

turns out to be about 3.88 percent. (This means that George can wait before buying houses for the yellow set.) Slightly more ambitiously, we can ask for the probability of *LandsOnYellowSet* given that the *second* roll is a double-3. The algorithm works out for itself that, in that case, the first roll must have been a double and concludes that the answer is about 36.1 percent. This is an example of Bayesian updating: when the new evidence (that the second roll is a double-3) is added, the probability of *LandsOnYellowSet* changes from 3.88 percent to 36.1

percent. Similarly, the probability that I roll three times ($Doubles_{34}$ is true) is 2.78 percent, while the probability that I roll three times given that I land on the yellow set is 20.44 percent.

Bayesian networks provide a way to build knowledge-based systems that avoids the failures that plagued the rule-based expert systems of the 1980s. (Indeed, had the AI community been less resistant to probability in the early 1980s, it might have avoided the AI winter that followed the rule-based expert system bubble.) Thousands of applications have been fielded, in areas ranging from medical diagnosis to terrorism prevention.[3]

Bayesian networks provide machinery for representing the necessary probabilities and performing the calculations to implement Bayesian updating for many complex tasks. Like propositional logic, however, they are quite limited in their ability to represent general knowledge. In many applications, the Bayesian network representation becomes very large and repetitive—for example, just as the rules of Go have to be repeated for every square in propositional logic, the probability-based rules of Monopoly have to be repeated for every player, for every location a player might be on, and for every move in the game. Such huge networks are virtually impossible to create by hand; instead, one would have to resort to code written in a traditional language such as C++ to generate and piece together multiple Bayes net fragments. While this is practical as an engineering solution for a specific problem, it is an obstacle to generality because the C++ code has to be written anew by a human expert for each application.

First-order probabilistic languages

It turns out, fortunately, that we can combine the expressiveness of first-order logic with the ability of Bayesian networks to capture probabilistic information concisely. This combination gives us the best of both worlds: *probabilistic* knowledge-based systems are able to handle a much wider range of real-world situations than either logical methods or Bayesian networks. For example, we can easily capture probabilistic knowledge about genetic inheritance:

> **for all** persons c, f, and m,
>> **if** f is the father of c **and** m is the mother of c
>>> **and** both f and m have blood type AB,
>> **then** c has blood type AB with probability 0.5.

The combination of first-order logic and probability actually gives us much more than just a way to express uncertain information about lots of objects. The reason is that when we add uncertainty to worlds containing objects, we get two new kinds of uncertainty: not just uncertainty about which facts are true or false but also uncertainty about what objects exist and uncertainty about which objects are which. These kinds of uncertainty are completely pervasive. The world does not come with a list of characters, like a Victorian play; instead, you gradually learn about the existence of objects from observation.

Sometimes the knowledge of new objects can be fairly definite, as when you open your hotel window and see the basilica of Sacré-Cœur for the first time; or it can be quite indefinite, as when you feel a gentle rumble that might be an earthquake or a passing subway train. And while the identity of Sacré-Cœur is quite unambiguous, the identity of subway trains is not: you might ride the same physical train hundreds of times without ever realizing it's the same one. Sometimes we don't need to resolve the uncertainty: I don't usually name all the tomatoes in a bag of cherry tomatoes and keep track of how well each one is doing, unless perhaps I am recording the progress of a tomato putrefaction experiment. For a class full of graduate students, on the other hand, I try my best to keep track of their identities. (Once, there were two research assistants in my group who had the same first and last names and were of very similar appearance and worked on closely related topics; at least, I am fairly sure there were two.) The problem is that we directly perceive not the *identity* of objects but (aspects of) their *appearance*; objects do not usually have little license plates that uniquely identify them. Identity is something our minds sometimes attach to objects for our own purposes.

The combination of probability theory with an expressive formal language is a fairly new subfield of AI, often called *probabilistic programming*.[4] Several dozen probabilistic programming languages, or PPLs, have been developed, many of them deriving their expressive power from ordinary programming languages rather than first-order logic. All PPL systems have the capacity to represent and reason with complex, uncertain knowledge. Applications include Microsoft's TrueSkill system, which rates millions of video game players every day; models for aspects of human cognition that were previously inexplicable by any mechanistic hypothesis, such as the ability to learn new visual categories of objects from single examples;[5] and the global seismic monitoring for the Comprehensive Nuclear-Test-Ban Treaty (CTBT), which is responsible for detecting clandestine nuclear explosions.[6]

The CTBT monitoring system collects real-time ground movement data from a global network of over 150 seismometers and aims to identify all the seismic events occurring on Earth above a certain magnitude and to flag the suspicious ones. Clearly there is plenty of existence uncertainty in this problem, because we don't know in advance the events that will occur; moreover, the vast majority of signals in the data are just noise. There is also lots of identity uncertainty: a blip of seismic energy detected at station A in Antarctica may or may not come from the same event as another blip detected at station B in Brazil. Listening to the Earth is like listening to thousands of simultaneous conversations that have been scrambled by transmission delays and echoes and drowned out by crashing waves.

How do we solve this problem using probabilistic programming? One might think we need some very clever algorithms to sort out all the possibilities. In fact, by following the methodology of knowledge-based systems, we don't have to devise any new algorithms at all. We

FIGURE 19: Location estimates for the February 12, 2013, nuclear test carried out by the government of North Korea. The tunnel entrance (black cross at lower center) was identified in satellite photographs. The NET-VISA location estimate is approximately 700 meters from the tunnel entrance and is based primarily on detections at stations 4,000 to 10,000 kilometers away. The CTBTO LEB location is the consensus estimate from expert geophysicists.

simply use a PPL to express what we know of geophysics: how often events tend to occur in areas of natural seismicity, how fast seismic waves travel through the Earth and how quickly they decay, how sensitive the detectors are, and how much noise there is. Then we add the data and run a probabilistic reasoning algorithm. The resulting monitoring system, called NET-VISA, has been operating as part of the treaty verification regime since 2018. Figure 19 shows NET-VISA's detection of a 2013 nuclear test in North Korea.

Keeping track of the world

One of the most important roles for probabilistic reasoning is in keeping track of parts of the world that are not directly observable. In

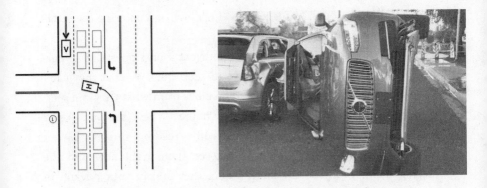

FIGURE 20: (left) Diagram of the situation leading up to the accident. The self-driving Volvo, marked V, is approaching an intersection, driving in the rightmost lane at thirty-eight miles per hour. Traffic in the other two lanes is stopped and the traffic light (L) is turning yellow. Invisible to the Volvo, a Honda (H) is making a left turn; (right) aftermath of the accident.

most video and board games, this is unnecessary because all the relevant information is observable, but in the real world this is seldom the case.

An example is given by one of the first serious accidents involving a self-driving car. It occurred on South McClintock Drive at East Don Carlos Avenue in Tempe, Arizona, on March 24, 2017.[7] As shown in figure 20, a self-driving Volvo (V), going south on McClintock, is approaching an intersection where the traffic light is just turning yellow. The Volvo's lane is clear, so it proceeds at the same speed through the intersection. Then a currently invisible vehicle—the Honda (H) in figure 20—appears from behind the queue of stopped traffic and a collision ensues.

To infer the possible presence of the invisible Honda, the Volvo could gather clues as it approaches the intersection. In particular, the traffic in the other two lanes is stopped even though the light is green; the cars at the front of the queue are not inching forward into the intersection and have their brake lights on. This is not *conclusive* evidence of an invisible left turner but it doesn't need to be; even a small

probability is enough to suggest slowing down and entering the intersection more cautiously.

The moral of this story is that intelligent agents operating in partially observable environments have to keep track of what they can't
see—to the extent possible—based on clues from what they can see.

Here's another example closer to home: Where are your keys?
Unless you happen to be driving while reading this book—not
recommended—you probably cannot see them right now. On the
other hand, you probably know where they are: in your pocket, in
your bag, on the bedside table, in the pocket of your coat which is
hanging up, or maybe on the hook in the kitchen. You know this because you put them there and they haven't moved since. This is a
simple example of using knowledge and reasoning to keep track of the
state of the world.

Without this capability, we would be lost—often quite literally. For
example, as I write this, I am looking at the white wall of a nondescript
hotel room. Where am I? If I had to rely on my current perceptual input, I would indeed be lost. In fact, I know that I am in Zürich, because
I arrived in Zürich yesterday and I haven't left. Like humans, robots
need to know where they are so that they can navigate successfully
through rooms, buildings, streets, forests, and deserts.

In AI we use the term *belief state* to refer to an agent's current
knowledge of the state of the world—however incomplete and uncertain it may be. Generally, the belief state—rather than the current
perceptual input—is the proper basis for making decisions about what
to do. Keeping the belief state up to date is a core activity for any intelligent agent. For some parts of the belief state, this happens automatically—for example, I just seem to know that I'm in Zürich,
without having to think about it. For other parts, it happens on demand, so to speak. For example, when I wake up in a new city with
severe jet lag, halfway through a long trip, I may have to make a conscious effort to reconstruct where I am, what I am supposed to be

doing, and why—a bit like a laptop rebooting itself, I suppose. Keeping track doesn't mean always knowing *exactly* the state of *everything* in the world. Obviously this is impossible—for example, I have no idea who is occupying the other rooms in my nondescript hotel in Zürich, let alone the present locations and activities of most of the eight billion people on Earth. I haven't the faintest idea what's happening in the rest of the universe beyond the solar system. My uncertainty about the current state of affairs is both massive and inevitable.

The basic method for keeping track of an uncertain world is *Bayesian updating.* Algorithms for doing this usually have two steps: a prediction step, where the agent predicts the current state of the world given its most recent action, and then an update step, where it receives new perceptual input and updates its beliefs accordingly. To illustrate how this works, consider the problem a robot faces in figuring out where it is. Figure 21(a) illustrates a typical case: The robot is in the middle of a room, with some uncertainty about its exact location, and wants to go through the door. It commands its wheels to move 1.5 meters towards the door; unfortunately, its wheels are old and wobbly, so the robot's prediction about where it ends up is quite uncertain, as shown in figure 21(b). If it tried to keep moving now, it might well crash. Fortunately, it has a sonar device to measure the distance to the doorposts. As figure 21(c) shows, the measurements suggest the robot is about 70 centimeters from the left doorpost and 85 centimeters from the right. Finally, the robot updates its belief state by combining the prediction in (b) with the measurements in (c) to obtain the new belief state in figure 21(d).

The algorithm for keeping track of the belief state can be applied to handle not just uncertainty about location but also uncertainty about the map itself. This results in a technique called SLAM (simultaneous localization and mapping). SLAM is a core component of many AI applications, ranging from augmented reality systems to self-driving cars and planetary rovers.

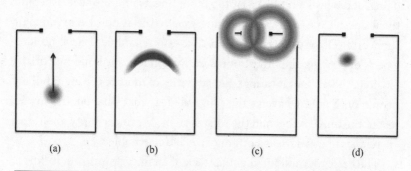

(a)   (b)   (c)   (d)

FIGURE 21: A robot trying to move through a doorway. (a) The initial belief state: the robot is somewhat uncertain of its location; it tries to move 1.5 meters towards the door. (b) The prediction step: the robot estimates that it is closer to the door but is quite uncertain about the direction it actually moved because its motors are old and its wheels wobbly. (c) The robot measures the distance to each doorpost using a poor-quality sonar device; the estimates are 70 centimeters from the left doorpost and 85 centimeters from the right. (d) The update step: combining the prediction in (b) with the observation in (c) gives the new belief state. Now the robot has a pretty good idea of where it is and will need to correct its course a bit to get through the door.

LEARNING FROM EXPERIENCE

L earning means improving performance based on experience. For a visual perception system, that might mean learning to recognize more categories of objects based on seeing examples of those categories; for a knowledge-based system, simply acquiring more knowledge is a form of learning, because it means the system can answer more questions; for a lookahead decision-making system such as AlphaGo, learning could mean improving its ability to evaluate positions or improving its ability to explore useful parts of the tree of possibilities.

Learning from examples

The most common form of machine learning is called *supervised* learning. A supervised learning algorithm is given a collection of training examples, each labeled with the correct output, and must produce a hypothesis as to what the correct rule is. Typically, a supervised learning system seeks to optimize the agreement between the hypothesis and the training examples. Often there is also a penalty for hypotheses that are more complicated than necessary—as recommended by Ockham's razor.

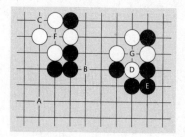

FIGURE 22: Legal and illegal moves in Go: moves A, B, and C are legal for Black, while moves D, E, and F are illegal. Move G might or might not be legal, depending on what has happened previously in the game.

Let's illustrate this for the problem of learning the legal moves in Go. (If you already know the rules of Go, then at least this will be easy to follow; if not, then you'll be better able to sympathize with the learning program.) Suppose the algorithm starts with the hypothesis

> **for all** time steps t, and **for all** locations l,
>> it is legal to play a stone at location l at time t.

It is Black's turn to move in the position shown in figure 22. The algorithm tries A: that's fine. B and C too. Then it tries D, on top of an existing white piece: that's illegal. (In chess or backgammon, it would be fine—that's how pieces are captured.) The move at E, on top of a black piece, is also illegal. (Illegal in chess too, but legal in backgammon.) Now, from these five training examples, the algorithm might propose the following hypothesis:

> **for all** time steps t, and **for all** locations l,
>> **if** l is unoccupied at time t,
>>> **then** it is legal to play a stone at location l at time t.

Then it tries F and finds to its surprise that F is illegal. After a few false starts, it settles on the following:

for all time steps *t*, and **for all** locations *l*,
  if *l* is unoccupied at time *t* **and**
    l is not surrounded by opponent stones,
  then it is legal to play a stone at location *l* at time *t*.

(This is sometimes called the *no suicide* rule.) Finally, it tries G, which in this case turns out to be legal. After scratching its head for a while and perhaps trying a few more experiments, it settles on the hypothesis that G is OK, even though it is surrounded, because it captures the white stone at D and therefore becomes un-surrounded immediately.

As you can see from the gradual progression of rules, learning takes place by a sequence of modifications to the hypothesis so as to fit the observed examples. This is something a learning algorithm can do easily. Machine learning researchers have designed all sorts of ingenious algorithms for finding good hypotheses quickly. Here the algorithm is searching in the space of logical expressions representing Go rules, but the hypotheses could also be algebraic expressions representing physical laws, probabilistic Bayesian networks representing diseases and symptoms, or even computer programs representing the complicated behavior of some other machine.

A second important point is that *even good hypotheses can be wrong*: in fact, the hypothesis given above *is* wrong, even after fixing it to ensure that G is legal. It needs to include the *ko* or *no-repetition* rule— for example, if White had just captured a black stone at G by playing at D, Black may not recapture by playing at G, since that produces the same position again. Notice that this rule is a radical departure from what the program has learned so far, because it means that legality cannot be determined from the current position; instead, one also has to remember previous positions.

The Scottish philosopher David Hume pointed out in 1748 that inductive reasoning—that is, reasoning from particular observations to general principles—can never be guaranteed.[1] In the modern theory of statistical learning, we ask not for guarantees of perfect correctness but only for a guarantee that the hypothesis found is *probably approximately correct*.[2] A learning algorithm can be "unlucky" and see an unrepresentative sample—for example, it might never try a move like G, thinking it to be illegal. It can also fail to predict some weird edge cases, such as the ones covered by some of the more complicated and rarely invoked forms of the no-repetition rule.[3] But, as long as the universe exhibits some degree of regularity, it's very unlikely that the algorithm could produce a seriously bad hypothesis, because such a hypothesis would very probably have been "found out" by one of the experiments.

Deep learning—the technology causing all the hullabaloo about AI in the media—is primarily a form of supervised learning. It represents one of the most significant advances in AI in recent decades, so it's worth understanding how it works. Moreover, some researchers believe it will lead to human-level AI systems within a few years, so it's a good idea to assess whether that's likely to be true.

It's easiest to understand deep learning in the context of a particular task, such as learning to distinguish giraffes and llamas. Given some labeled photographs of each, the learning algorithm has to form a hypothesis that allows it to classify unlabeled images. An image is, from the computer's point of view, nothing but a large table of numbers, with each number corresponding to one of three RGB values for one pixel of the image. So, instead of a Go hypothesis that takes a board position and a move as input and decides whether the move is legal, we need a giraffe–llama hypothesis that takes a table of numbers as input and predicts a category (giraffe or llama).

Now the question is, what sort of hypothesis? Over the last fifty-odd years of computer vision research, many approaches have been tried. The current favorite is a *deep convolutional network*. Let me unpack this: It's called a *network* because it represents a complex mathematical expression composed in a regular way from many smaller subexpressions, and the compositional structure has the form of a network. (Such networks are often called *neural networks* because their designers draw inspiration from the networks of neurons in the brain.) It's called *convolutional* because that's a fancy mathematical way to say that the network structure repeats itself in a fixed pattern across the whole input image. And it's called *deep* because such networks typically have many layers, and also because it sounds impressive and slightly spooky.

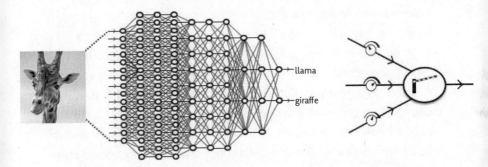

FIGURE 23: (left) A simplified depiction of a deep convolutional network for recognizing objects in images. The image pixel values are fed in at the left and the network outputs values at the two rightmost nodes, indicating how likely the image is to be a llama or a giraffe. Notice how the pattern of local connections, indicated by the dark lines in the first layer, repeats across the whole layer; (right) one of the nodes in the network. There is an adjustable weight on each incoming value so that the node pays more or less attention to it. Then the total incoming signal goes through a gating function that allows large signals through but suppresses small ones.

A simplified example (simplified because real networks may have hundreds of layers and millions of nodes) is shown in figure 23. The network is really a picture of a complex, adjustable mathematical expression. Each node in the network corresponds to a simple adjustable expression, as illustrated in the figure. Adjustments are made by changing the *weights* on each input, as indicated by the "volume controls." The weighted sum of the inputs is then passed through a gating function before reaching the output side of the node; typically, the gating function suppresses small values and allows larger ones through.

Learning takes place in the network simply by adjusting all the volume control knobs to reduce the prediction error on the labeled examples. It's as simple as that: no magic, no especially ingenious algorithms. Working out which way to turn the knobs to decrease the error is a straightforward application of calculus to compute how changing each weight would change the error at the output layer. This leads to a simple formula for propagating the error backwards from the output layer to the input layer, tweaking knobs along the way.

Miraculously, the process works. For the task of recognizing objects in photographs, deep learning algorithms have demonstrated remarkable performance. The first inkling of this came in the 2012 ImageNet competition, which provides training data consisting of 1.2 million labeled images in one thousand categories, and then requires the algorithm to label one hundred thousand new images.[4] Geoff Hinton, a British computational psychologist who was at the forefront of the first neural network revolution in the 1980s, had been experimenting with a very large deep convolutional network: 650,000 nodes and 60 million parameters. He and his group at the University of Toronto achieved an ImageNet error rate of 15 percent, a dramatic improvement on the previous best of 26 percent.[5] By 2015, dozens of teams were using deep learning methods and the error rate was down to 5 percent, comparable to that of a human who had spent weeks learning to recognize the thousand categories in the test.[6] By 2017, the machine error rate was 2 percent.

Over roughly the same period, there have been comparable improvements in speech recognition and machine translation based on similar methods. Taken together, these are three of the most important application areas for AI. Deep learning has also played an important role in applications of reinforcement learning—for example, in learning the evaluation function that AlphaGo uses to estimate the desirability of possible future positions, and in learning controllers for complex robotic behaviors.

As yet, we have very little understanding as to why deep learning works as well as it does. Possibly the best explanation is that deep networks are deep: because they have many layers, each layer can learn a fairly simple transformation from its inputs to its outputs, while many such simple transformations add up to the complex transformation required to go from a photograph to a category label. In addition, deep networks for vision have built-in structure that enforces translation invariance and scale invariance—meaning that a dog is a dog no matter where it appears in the image and no matter how big it appears in the image.

Another important property of deep networks is that they often seem to discover internal representations that capture elementary features of images, such as eyes, stripes, and simple shapes. None of these features are built in. We know they are there because we can experiment with the trained network and see what kinds of data cause the internal nodes (typically those close to the output layer) to light up. In fact, it is possible to run the learning algorithm a different way so that it adjusts the image itself to produce a stronger response at chosen internal nodes. Repeating this process many times produces what are now known as *deep dreaming* or *inceptionism* images, such as the one in figure 24.[7] Inceptionism has become an art form in itself, producing images unlike any human art.

For all their remarkable achievements, deep learning systems as we currently understand them are far from providing a basis for generally intelligent systems. Their principal weakness is that they are *circuits*; they are cousins of propositional logic and Bayesian networks, which, for all their wonderful properties, also lack the ability to express complex forms of knowledge in a concise way. This means that deep networks operating in "native mode" require vast amounts of circuitry

FIGURE 24: An image generated by Google's DeepDream software.

to represent fairly simple kinds of general knowledge. That, in turn, implies vast numbers of weights to learn and hence a need for unreasonable numbers of examples—more than the universe could ever supply.

Some argue that the brain is also made of circuits, with neurons as the circuit elements; therefore, circuits can support human-level intelligence. This is true, but only in the same sense that brains are made of atoms: atoms can indeed support human-level intelligence, but that doesn't mean that just collecting together lots of atoms will produce intelligence. The atoms have to be arranged in certain ways. By the same token, the circuits have to be arranged in certain ways. Computers are also made of circuits, both in their memories and in their processing units; but those circuits have to be arranged in certain ways, and layers of software have to be added, before the computer can support the operation of high-level programming languages and logical reasoning systems. At present, however, there is no sign that deep learning systems can develop such capabilities by themselves— nor does it make scientific sense to require them to do so.

There are further reasons to think that deep learning may reach a plateau well short of general intelligence, but it's not my purpose here to diagnose all the problems: others, both inside[8] and outside[9] the deep learning community, have noted many of them. The point is that simply creating larger and deeper networks and larger data sets and bigger machines is not enough to create human-level AI. We have already seen (in Appendix B) DeepMind CEO Demis Hassabis's view that "higher-level thinking and symbolic reasoning" are essential for AI. Another prominent deep learning expert, François Chollet, put it this way:[10] "Many more applications are completely out of reach for current deep learning techniques—even given vast amounts of human-annotated data. . . . We need to move away from straightforward input-to-output mappings, and on to reasoning and abstraction."

Learning from thinking

Whenever you find yourself having to think about something, it's because you don't already know the answer. When someone asks for the number of your brand-new cell phone, you probably don't know it. You think to yourself, "OK, I don't know it; so how do I find it?" Not being a slave to the cell phone, you don't know how to find it. You think to yourself, "How do I figure out how to find it?" You have a generic answer to this: "Probably they put it somewhere that's easy for users to find." (Of course, you could be wrong about this.) Obvious places would be at the top of the home screen (not there), inside the Phone app, or in Settings for that app. You try Settings>Phone, and there it is.

The next time you are asked for your number, you either know it or you know exactly how to get it. You remember the procedure, not just for *this* phone on *this* occasion but for *all* similar phones on *all* occasions—that is, you store and reuse a *generalized* solution to the problem. The generalization is justified because you understand that the specifics of this particular phone and this particular occasion are irrelevant. You would be shocked if the method worked only on Tuesdays for phone numbers ending in 17.

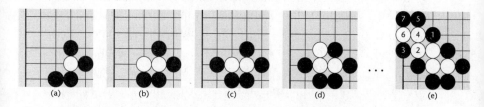

FIGURE 25: The concept of a *ladder* in Go. (a) Black threatens to capture White's piece. (b) White tries to escape. (c) Black blocks that direction of escape. (d) White tries the other direction. (e) Play continues in the sequence indicated by the numbers. The ladder eventually reaches the edge of the board, where White has nowhere to run. The coup de grâce is administered by move 7: White's group is completely surrounded and dies.

Go offers a beautiful example of the same kind of learning. In figure 25(a), we see a common situation where Black threatens to capture White's stone by surrounding it. White attempts to escape by adding stones connected to the original one, but Black continues to cut off the routes of escape. This pattern of moves forms a *ladder* of stones diagonally across the board, until it runs into the edge; then White has nowhere to go. If you are White, you probably won't make the same mistake again: you realize that the ladder pattern *always* results in eventual capture, for *any* initial location and *any* direction, at *any* stage of the game, whether you are playing White or Black. The only exception occurs when the ladder runs into some additional stones belonging to the escapee. The generality of the ladder pattern follows straightforwardly from the rules of Go.

The case of the missing phone number and the case of the Go ladder illustrate the possibility of learning effective, general rules from a single example—a far cry from the millions of examples needed for deep learning. In AI, this kind of learning is called *explanation-based learning*: on seeing the example, the agent can explain to itself *why* it came out that way and can extract the general principle by seeing what factors were essential for the explanation.

Strictly speaking, the process does not, by itself, add new knowledge—for example, White could have simply derived the existence and outcome of the general ladder pattern from the rules of Go, without ever seeing an example.[11] Chances are, however, that White wouldn't ever discover the ladder concept without seeing an example of it; so, we can understand explanation-based learning as a powerful method for saving the results of computation in a generalized way, so as to avoid having to recapitulate the same reasoning process (or making the same mistake with an imperfect reasoning process) in the future.

Research in cognitive science has stressed the importance of this type of learning in human cognition. Under the name of *chunking*, it forms a central pillar of Allen Newell's highly influential theory of cognition.[12] (Newell was one of the attendees of the 1956 Dartmouth workshop and co-winner of the 1975 Turing Award with Herb Simon.) It explains how humans become more fluent at cognitive tasks with practice, as various subtasks that originally required thinking become automatic. Without it, human conversations would be limited to one- or two-word responses and mathematicians would still be counting on their fingers.

Acknowledgments

Many people have helped in the creation of this book. They include my excellent editors at Viking (Paul Slovak) and Penguin (Laura Stickney); my agent, John Brockman, who encouraged me to write something; Jill Leovy and Rob Reid, who provided reams of useful feedback; and other readers of early drafts, especially Ziyad Marar, Nick Hay, Toby Ord, David Duvenaud, Max Tegmark, and Grace Cassy. Caroline Jeanmaire was immensely helpful in collating the innumerable suggestions for improvements made by the early readers, and Martin Fukui handled the collecting of permissions for images.

The main technical ideas in the book have been developed in collaboration with the members of the Center for Human-Compatible AI at Berkeley, especially Tom Griffiths, Anca Dragan, Andrew Critch, Dylan Hadfield-Menell, Rohin Shah, and Smitha Milli. The Center has been admirably piloted by executive director Mark Nitzberg and assistant director Rosie Campbell, and generously funded by the Open Philanthropy Foundation.

Ramona Alvarez and Carine Verdeau helped to keep things running throughout the process, and my incredible wife, Loy, and our children—Gordon, Lucy, George, and Isaac—supplied copious and necessary amounts of love, forbearance, and encouragement to finish, not always in that order.

Notes

CHAPTER 1

1. The first edition of my textbook on AI, co-authored with Peter Norvig, currently director of research at Google: Stuart Russell and Peter Norvig, *Artificial Intelligence: A Modern Approach*, 1st ed. (Prentice Hall, 1995).
2. Robinson developed the *resolution* algorithm, which can, given enough time, prove any logical consequence of a set of first-order logical assertions. Unlike previous algorithms, it did not require conversion to propositional logic. J. Alan Robinson, "A machine-oriented logic based on the resolution principle," *Journal of the ACM* 12 (1965): 23–41.
3. Arthur Samuel, an American pioneer of the computer era, did his early work at IBM. The paper describing his work on checkers was the first to use the term *machine learning*, although Alan Turing had already talked about "a machine that can learn from experience" as early as 1947. Arthur Samuel, "Some studies in machine learning using the game of checkers," *IBM Journal of Research and Development* 3 (1959): 210–29.
4. The "Lighthill Report," as it became known, led to the termination of research funding for AI except at the universities of Edinburgh and Sussex: Michael James Lighthill, "Artificial intelligence: A general survey," in *Artificial Intelligence: A Paper Symposium* (Science Research Council of Great Britain, 1973).
5. The CDC 6600 filled an entire room and cost the equivalent of $20 million. For its era it was incredibly powerful, albeit a million times less powerful than an iPhone.
6. Following Deep Blue's victory over Kasparov, at least one commentator predicted that it would take one hundred years before the same thing happened in Go: George Johnson, "To test a powerful computer, play an ancient game," *The New York Times*, July 29, 1997.
7. For a highly readable history of the development of nuclear technology, see Richard Rhodes, *The Making of the Atomic Bomb* (Simon & Schuster, 1987).
8. A simple supervised learning algorithm may not have this effect, unless it is wrapped within an A/B testing framework (as is common in online marketing settings). Bandit algorithms and reinforcement learning algorithms will have this effect if they operate with an explicit representation of user state or an implicit representation in terms of the history of interactions with the user.
9. Some have argued that profit-maximizing corporations are already out-of-control artificial entities. See, for example, Charles Stross, "Dude, you broke the future!" (keynote, 34th Chaos Communications Congress, 2017). See also Ted Chiang, "Silicon Valley is turning into its own worst fear," *Buzzfeed*, December 18, 2017. The idea is explored further by Daniel Hillis, "The first machine intelligences," in *Possible Minds: Twenty-Five Ways of Looking at AI*, ed. John Brockman (Penguin Press, 2019).

10. For its time, Wiener's paper was a rare exception to the prevailing view that all tech-
    nological progress was a good thing: Norbert Wiener, "Some moral and technical con-
    sequences of automation," *Science* 131 (1960): 1355–58.

CHAPTER 2

 1. Santiago Ramón y Cajal proposed synaptic changes as the site of learning in 1894, but
    it was not until the late 1960s that this hypothesis was confirmed experimentally. See
    Timothy Bliss and Terje Lomo, "Long-lasting potentiation of synaptic transmission in
    the dentate area of the anaesthetized rabbit following stimulation of the perforant
    path," *Journal of Physiology* 232 (1973): 331–56.
 2. For a brief introduction, see James Gorman, "Learning how little we know about the
    brain," *The New York Times*, November 10, 2014. See also Tom Siegfried, "There's a
    long way to go in understanding the brain," *ScienceNews*, July 25, 2017. A special 2017
    issue of the journal *Neuron* (vol. 94, pp. 933–1040) provides a good overview of many
    different approaches to understanding the brain.
 3. The presence or absence of consciousness—actual subjective experience—certainly
    makes a difference in our moral consideration for machines. If ever we gain enough
    understanding to design conscious machines or to detect that we have done so, we
    would face many important moral issues for which we are largely unprepared.
 4. The following paper was among the first to make a clear connection between re-
    inforcement learning algorithms and neurophysiological recordings: Wolfram Schultz,
    Peter Dayan, and P. Read Montague, "A neural substrate of prediction and reward,"
    Science 275 (1997): 1593–99.
 5. Studies of intracranial stimulation were carried out with the hope of finding cures for
    various mental illnesses. See, for example, Robert Heath, "Electrical self-stimulation
    of the brain in man," *American Journal of Psychiatry* 120 (1963): 571–77.
 6. An example of a species that may be facing self-extinction via addiction: Bryson Voi-
    rin, "Biology and conservation of the pygmy sloth, *Bradypus pygmaeus*," *Journal of
    Mammalogy* 96 (2015): 703–7.
 7. The *Baldwin effect* in evolution is usually attributed to the following paper: James
    Baldwin, "A new factor in evolution," *American Naturalist* 30 (1896): 441–51.
 8. The core idea of the Baldwin effect also appears in the following work: Conwy Lloyd
    Morgan, *Habit and Instinct* (Edward Arnold, 1896).
 9. A modern analysis and computer implementation demonstrating the Baldwin effect:
    Geoffrey Hinton and Steven Nowlan, "How learning can guide evolution," *Complex
    Systems* 1 (1987): 495–502.
10. Further elucidation of the Baldwin effect by a computer model that includes the evo-
    lution of the internal reward-signaling circuitry: David Ackley and Michael Littman,
    "Interactions between learning and evolution," in *Artificial Life II*, ed. Christopher
    Langton et al. (Addison-Wesley, 1991).
11. Here I am pointing to the roots of our present-day concept of intelligence, rather than
    describing the ancient Greek concept of *nous*, which had a variety of related meanings.
12. The quotation is taken from Aristotle, *Nicomachean Ethics*, Book III, 3, 1112b.
13. Cardano, one of the first European mathematicians to consider negative numbers,
    developed an early mathematical treatment of probability in games. He died in 1576,
    eighty-seven years before his work appeared in print: Gerolamo Cardano, *Liber de ludo
    aleae* (Lyons, 1663).
14. Arnauld's work, initially published anonymously, is often called *The Port-Royal Logic*:
    Antoine Arnauld, *La logique, ou l'art de penser* (Chez Charles Savreux, 1662). See also
    Blaise Pascal, *Pensées* (Chez Guillaume Desprez, 1670).
15. The concept of utility: Daniel Bernoulli, "Specimen theoriae novae de mensura sortis,"
    Proceedings of the St. Petersburg Imperial Academy of Sciences 5 (1738): 175–92. Bernoul-
    li's idea of utility arises from considering a merchant, Sempronius, choosing whether to
    transport a valuable cargo in one ship or to split it between two, assuming that each ship
    has a 50 percent probability of sinking on the journey. The expected monetary value of
    the two solutions is the same, but Sempronius clearly prefers the two-ship solution.

16. By most accounts, von Neumann did not himself invent this architecture but his name was on an early draft of an influential report describing the EDVAC stored-program computer.
17. The work of von Neumann and Morgenstern is in many ways the foundation of modern economic theory: John von Neumann and Oskar Morgenstern, *Theory of Games and Economic Behavior* (Princeton University Press, 1944).
18. The proposal that utility is a sum of discounted rewards was put forward as a mathematically convenient hypothesis by Paul Samuelson, "A note on measurement of utility," *Review of Economic Studies* 4 (1937): 155–61. If s_0, s_1, . . . is a sequence of states, then its utility in this model is $U(s_0, s_1, . . .) = \Sigma_t \gamma^t R(s_t)$, where γ is a discount factor and R is a reward function describing the desirability of a state. Naïve application of this model seldom agrees with the judgment of real individuals about the desirability of present and future rewards. For a thorough analysis, see Shane Frederick, George Loewenstein, and Ted O'Donoghue, "Time discounting and time preference: A critical review," *Journal of Economic Literature* 40 (2002): 351–401.
19. Maurice Allais, a French economist, proposed a decision scenario in which humans appear consistently to violate the von Neumann–Morgenstern axioms: Maurice Allais, "Le comportement de l'homme rationnel devant le risque: Critique des postulats et axiomes de l'école américaine," *Econometrica* 21 (1953): 503–46.
20. For an introduction to non-quantitative decision analysis, see Michael Wellman, "Fundamental concepts of qualitative probabilistic networks," *Artificial Intelligence* 44 (1990): 257–303.
21. I will discuss the evidence for human irrationality further in Chapter 9. The standard references include the following: Allais, "Le comportement"; Daniel Ellsberg, *Risk, Ambiguity, and Decision* (PhD thesis, Harvard University, 1962); Amos Tversky and Daniel Kahneman, "Judgment under uncertainty: Heuristics and biases," *Science* 185 (1974): 1124–31.
22. It should be clear that this is a thought experiment that cannot be realized in practice. Choices about different futures are never presented in full detail, and humans never have the luxury of minutely examining and savoring those futures before choosing. Instead, one is given only brief summaries, such as "librarian" or "coal miner." In making such a choice, one is really being asked to compare two probability distributions over complete futures, one beginning with the choice "librarian" and the other "coal miner," with each distribution assuming optimal actions on one's own part within each future. Needless to say, this is not easy.
23. The first mention of a randomized strategy for games appears in Pierre Rémond de Montmort, *Essay d'analyse sur les jeux de hazard*, 2nd ed. (Chez Jacques Quillau, 1713). The book identifies a certain Monsieur de Waldegrave as the source of an optimal randomized solution for the card game Le Her. Details of Waldegrave's identity are revealed by David Bellhouse, "The problem of Waldegrave," *Electronic Journal for History of Probability and Statistics* 3 (2007).
24. The problem is fully defined by specifying the probability that Alice scores in each of four cases: when she shoots to Bob's right and he dives right or left, and when she shoots to his left and he dives right or left. In this case, these probabilities are 25 percent, 70 percent, 65 percent, and 10 percent respectively. Now suppose that Alice's strategy is to shoot to Bob's right with probability p and his left with probability $1 - p$, while Bob dives to his right with probability q and left with probability $1 - q$. The payoff to Alice is $U_A = 0.25pq + 0.70\,p(1 - q) + 0.65\,(1 - p)q + 0.10(1 - p)\,(1 - q)$, while Bob's payoff is $U_B = -U_A$. At equilibrium, $\partial U_A/\partial p = 0$ and $\partial U_B/\partial q = 0$, giving $p = 0.55$ and $q = 0.60$.
25. The original game-theoretic problem was introduced by Merrill Flood and Melvin Dresher at the RAND Corporation; Tucker saw the payoff matrix on a visit to their offices and proposed a "story" to go along with it.
26. Game theorists typically say that Alice and Bob could *cooperate* with each other (refuse to talk) or *defect* and rat on their accomplice. I find this language confusing, because "cooperate with each other" is not a choice that each agent can make separately, and because in common parlance one often talks about cooperating with the police, receiving a lighter sentence in return for cooperating, and so on.

27. For an interesting trust-based solution to the prisoner's dilemma and other games, see Joshua Letchford, Vincent Conitzer, and Kamal Jain, "An 'ethical' game-theoretic solution concept for two-player perfect-information games," in *Proceedings of the 4th International Workshop on Web and Internet Economics*, ed. Christos Papadimitriou and Shuzhong Zhang (Springer, 2008).

28. Origin of the tragedy of the commons: William Forster Lloyd, *Two Lectures on the Checks to Population* (Oxford University, 1833).

29. Modern revival of the topic in the context of global ecology: Garrett Hardin, "The tragedy of the commons," *Science* 162 (1968): 1243–48.

30. It's quite possible that even if we had tried to build intelligent machines from chemical reactions or biological cells, those assemblages would have turned out to be implementations of Turing machines in nontraditional materials. Whether an object is a general-purpose computer has nothing to do with what it's made of.

31. Turing's breakthrough paper defined what is now known as the *Turing machine*, the basis for modern computer science. The *Entscheidungsproblem*, or *decision problem*, in the title is the problem of deciding entailment in first-order logic: Alan Turing, "On computable numbers, with an application to the *Entscheidungsproblem*," *Proceedings of the London Mathematical Society*, 2nd ser., 42 (1936): 230–65.

32. A good survey of research on negative capacitance by one of its inventors: Sayeef Salahuddin, "Review of negative capacitance transistors," in *International Symposium on VLSI Technology, Systems and Application* (IEEE Press, 2016).

33. For a much better explanation of quantum computation, see Scott Aaronson, *Quantum Computing since Democritus* (Cambridge University Press, 2013).

34. The paper that established a clear complexity-theoretic distinction between classical and quantum computation: Ethan Bernstein and Umesh Vazirani, "Quantum complexity theory," *SIAM Journal on Computing* 26 (1997): 1411–73.

35. The following article by a renowned physicist provides a good introduction to the current state of understanding and technology: John Preskill, "Quantum computing in the NISQ era and beyond," arXiv:1801.00862 (2018).

36. On the maximum computational ability of a one-kilogram object: Seth Lloyd, "Ultimate physical limits to computation," *Nature* 406 (2000): 1047–54.

37. For an example of the suggestion that humans may be the pinnacle of physically achievable intelligence, see Kevin Kelly, "The myth of a superhuman AI," *Wired*, April 25, 2017: "We tend to believe that the limit is way beyond us, way 'above' us, as we are 'above' an ant. . . . What evidence do we have that the limit is not us?"

38. In case you are wondering about a simple trick to solve the halting problem: the obvious method of just running the program to see if it finishes doesn't work, because that method doesn't necessarily finish. You might wait a million years and still not know if the program is really stuck in an infinite loop or just taking its time.

39. The proof that the halting problem is undecidable is an elegant piece of trickery. The question: Is there a LoopChecker(P,X) program that, for *any* program P and *any* input X, decides correctly, in finite time, whether P applied to input X will halt and produce a result or keep chugging away forever? Suppose that LoopChecker exists. Now write a program Q that calls LoopChecker as a subroutine, with Q itself and X as inputs, and then does the *opposite* of what LoopChecker(Q,X) predicts. So, if LoopChecker says that Q halts, Q doesn't halt, and vice versa. Thus, the assumption that LoopChecker exists leads to a contradiction, so LoopChecker cannot exist.

40. I say "appear" because, as yet, the claim that the class of NP-complete problems requires superpolynomial time (usually referred to as P ≠ NP) is still an unproven conjecture. After almost fifty years of research, however, nearly all mathematicians and computer scientists are convinced the claim is true.

41. Lovelace's writings on computation appear mainly in her notes attached to her translation of an Italian engineer's commentary on Babbage's engine: L. F. Menabrea, "Sketch of the Analytical Engine invented by Charles Babbage," trans. Ada, Countess of Lovelace, in *Scientific Memoirs*, vol. III, ed. R. Taylor (R. and J. E. Taylor, 1843). Menabrea's original article, written in French and based on lectures given by Babbage in 1840, appears in *Bibliothèque Universelle de Genève* 82 (1842).

42. One of the seminal early papers on the possibility of artificial intelligence: Alan Turing, "Computing machinery and intelligence," *Mind* 59 (1950): 433–60.

43. The Shakey project at SRI is summarized in a retrospective by one of its leaders: Nils Nilsson, "Shakey the robot," technical note 323 (SRI International, 1984). A twenty-four-minute film, *SHAKEY: Experimentation in Robot Learning and Planning*, was made in 1969 and garnered national attention.

44. The book that marked the beginning of modern, probability-based AI: Judea Pearl, *Probabilistic Reasoning in Intelligent Systems: Networks of Plausible Inference* (Morgan Kaufmann, 1988).

45. Technically, chess is not fully observable. A program does need to remember a small amount of information to determine the legality of castling and en passant moves and to define draws by repetition or by the fifty-move rule.

46. For a complete exposition, see Chapter 2 of Stuart Russell and Peter Norvig, *Artificial Intelligence: A Modern Approach*, 3rd ed. (Pearson, 2010).

47. The size of the state space for StarCraft is discussed by Santiago Ontañon et al., "A survey of real-time strategy game AI research and competition in StarCraft," *IEEE Transactions on Computational Intelligence and AI in Games* 5 (2013): 293–311. Vast numbers of moves are possible because a player can move all units simultaneously. The numbers go down as restrictions are imposed on how many units or groups of units can be moved at once.

48. On human–machine competition in StarCraft: Tom Simonite, "DeepMind beats pros at StarCraft in another triumph for bots," *Wired*, January 25, 2019.

49. AlphaZero is described by David Silver et al., "Mastering chess and shogi by self-play with a general reinforcement learning algorithm," arXiv:1712.01815 (2017).

50. Optimal paths in graphs are found using the A* algorithm and its many descendants: Peter Hart, Nils Nilsson, and Bertram Raphael, "A formal basis for the heuristic determination of minimum cost paths," *IEEE Transactions on Systems Science and Cybernetics* SSC-4 (1968): 100–107.

51. The paper that introduced the Advice Taker program and logic-based knowledge systems: John McCarthy, "Programs with common sense," in *Proceedings of the Symposium on Mechanisation of Thought Processes* (Her Majesty's Stationery Office, 1958).

52. To get some sense of the significance of knowledge-based systems, consider database systems. A database contains concrete, individual facts, such as the location of my keys and the identities of your Facebook friends. Database systems cannot store general rules, such as the rules of chess or the legal definition of British citizenship. They can count how many people called Alice have friends called Bob, but they cannot determine whether a particular Alice meets the conditions for British citizenship or whether a particular sequence of moves on a chessboard will lead to checkmate. Database systems cannot combine two pieces of knowledge to produce a third: they support memory but not reasoning. (It is true that many modern database systems provide a way to add rules and a way to use those rules to derive new facts; to the extent that they do, they are really knowledge-based systems.) Despite being highly constricted versions of knowledge-based systems, database systems underlie most of present-day commercial activity and generate hundreds of billions of dollars in value every year.

53. The original paper describing the completeness theorem for first-order logic: Kurt Gödel, "Die Vollständigkeit der Axiome des logischen Funktionenkalküls," *Monatshefte für Mathematik* 37 (1930): 349–60.

54. The reasoning algorithm for first-order logic does have a gap: if there is no answer—that is, if the available knowledge is insufficient to give an answer either way—then the algorithm may never finish. This is unavoidable: it is mathematically *impossible* for a correct algorithm *always* to terminate with "don't know," for essentially the same reason that no algorithm can solve the halting problem (page 37).

55. The first algorithm for theorem-proving in first-order logic worked by reducing first-order sentences to (very large numbers of) propositional sentences: Martin Davis and Hilary Putnam, "A computing procedure for quantification theory," *Journal of the ACM* 7 (1960): 201–15. Robinson's resolution algorithm operated directly on first-order

logical sentences, using "unification" to match complex expressions containing logical variables: J. Alan Robinson, "A machine-oriented logic based on the resolution principle," *Journal of the ACM* 12 (1965): 23–41.

56. One might wonder how Shakey the logical robot ever reached any definite conclusions about what to do. The answer is simple: Shakey's knowledge base contained false assertions. For example, Shakey believed that by executing "push object A through door D into room B," object A would end up in room B. This belief was false because Shakey could get stuck in the doorway or miss the doorway altogether or someone might sneakily remove object A from Shakey's grasp. Shakey's plan execution module could detect plan failure and replan accordingly, so Shakey was not, strictly speaking, a purely logical system.

57. An early commentary on the role of probability in human thinking: Pierre-Simon Laplace, *Essai philosophique sur les probabilités* (Mme. Ve. Courcier, 1814).

58. Bayesian logic described in a fairly nontechnical way: Stuart Russell, "Unifying logic and probability," *Communications of the ACM* 58 (2015): 88–97. The paper draws heavily on the PhD thesis research of my former student Brian Milch.

59. The original source for Bayes' theorem: Thomas Bayes and Richard Price, "An essay towards solving a problem in the doctrine of chances," *Philosophical Transactions of the Royal Society of London* 53 (1763): 370–418.

60. Technically, Samuel's program did not treat winning and losing as absolute rewards; by fixing the value of material to be positive; however, the program generally tended to work towards winning.

61. The application of reinforcement learning to produce a world-class backgammon program: Gerald Tesauro, "Temporal difference learning and TD-Gammon," *Communications of the ACM* 38 (1995): 58–68.

62. The DQN system that learns to play a wide variety of video games using deep RL: Volodymyr Mnih et al., "Human-level control through deep reinforcement learning," *Nature* 518 (2015): 529–33.

63. Bill Gates's remarks on Dota 2 AI: Catherine Clifford, "Bill Gates says gamer bots from Elon Musk-backed nonprofit are 'huge milestone' in A.I.," *CNBC*, June 28, 2018.

64. An account of OpenAI Five's victory over the human world champions at Dota 2: Kelsey Piper, "AI triumphs against the world's top pro team in strategy game Dota 2," *Vox*, April 13, 2019.

65. A compendium of cases in the literature where misspecification of reward functions led to unexpected behavior: Victoria Krakovna, "Specification gaming examples in AI," *Deep Safety* (blog), April 2, 2018.

66. A case where an evolutionary fitness function defined in terms of maximum velocity led to very unexpected results: Karl Sims, "Evolving virtual creatures," in *Proceedings of the 21st Annual Conference on Computer Graphics and Interactive Techniques* (ACM, 1994).

67. For a fascinating exposition of the possibilities of reflex agents, see Valentino Braitenberg, *Vehicles: Experiments in Synthetic Psychology* (MIT Press, 1984).

68. News article on a fatal accident involving a vehicle in autonomous mode that hit a pedestrian: Devin Coldewey, "Uber in fatal crash detected pedestrian but had emergency braking disabled," *TechCrunch*, May 24, 2018.

69. On steering control algorithms, see, for example, Jarrod Snider, "Automatic steering methods for autonomous automobile path tracking," technical report CMU-RI-TR-09-08, Robotics Institute, Carnegie Mellon University, 2009.

70. Norfolk and Norwich terriers are two categories in the ImageNet database. They are notoriously hard to tell apart and were viewed as a single breed until 1964.

71. A very unfortunate incident with image labeling: Daniel Howley, "Google Photos mislabels 2 black Americans as gorillas," *Yahoo Tech*, June 29, 2015.

72. Follow-up article on Google and gorillas: Tom Simonite, "When it comes to gorillas, Google Photos remains blind," *Wired*, January 11, 2018.

CHAPTER 3

1. The basic plan for game-playing algorithms was laid out by Claude Shannon, "Programming a computer for playing chess," *Philosophical Magazine*, 7th ser., 41 (1950): 256–75.
2. See figure 5.12 of Stuart Russell and Peter Norvig, *Artificial Intelligence: A Modern Approach*, 1st ed. (Prentice Hall, 1995). Note that the rating of chess players and chess programs is not an exact science. Kasparov's highest-ever Elo rating was 2851, achieved in 1999, but current chess engines such as Stockfish are rated at 3300 or more.
3. The earliest reported autonomous vehicle on a public road: Ernst Dickmanns and Alfred Zapp, "Autonomous high speed road vehicle guidance by computer vision," *IFAC Proceedings Volumes* 20 (1987): 221–26.
4. The safety record for Google (subsequently Waymo) vehicles: "Waymo safety report: On the road to fully self-driving," 2018.
5. So far there have been at least two driver fatalities and one pedestrian fatality. Some references follow, along with brief quotes describing what happened. Danny Yadron and Dan Tynan, "Tesla driver dies in first fatal crash while using autopilot mode," *Guardian*, June 30, 2016: "The autopilot sensors on the Model S failed to distinguish a white tractor-trailer crossing the highway against a bright sky." Megan Rose Dickey, "Tesla Model X sped up in Autopilot mode seconds before fatal crash, according to NTSB," *TechCrunch*, June 7, 2018: "At 3 seconds prior to the crash and up to the time of impact with the crash attenuator, the Tesla's speed increased from 62 to 70.8 mph, with no precrash braking or evasive steering movement detected." Devin Coldewey, "Uber in fatal crash detected pedestrian but had emergency braking disabled," *TechCrunch*, May 24, 2018: "Emergency braking maneuvers are not enabled while the vehicle is under computer control, to reduce the potential for erratic vehicle behavior."
6. The Society of Automotive Engineers (SAE) defines six levels of automation, where Level 0 is none at all and Level 5 is full automation: "The full-time performance by an automatic driving system of all aspects of the dynamic driving task under all roadway and environmental conditions that can be managed by a human driver."
7. Forecast of economic effects of automation on transportation costs: Adele Peters, "It could be 10 times cheaper to take electric robo-taxis than to own a car by 2030," *Fast Company*, May 30, 2017.
8. The impact of accidents on the prospects for regulatory action on autonomous vehicles: Richard Waters, "Self-driving car death poses dilemma for regulators," *Financial Times*, March 20, 2018.
9. The impact of accidents on public perception of autonomous vehicles: Cox Automotive, "Autonomous vehicle awareness rising, acceptance declining, according to Cox Automotive mobility study," August 16, 2018.
10. The original chatbot: Joseph Weizenbaum, "ELIZA—a computer program for the study of natural language communication between man and machine," *Communications of the ACM* 9 (1966): 36–45.
11. See physiome.org for current activities in physiological modeling. Work in the 1960s assembled models with thousands of differential equations: Arthur Guyton, Thomas Coleman, and Harris Granger, "Circulation: Overall regulation," *Annual Review of Physiology* 34 (1972): 13–44.
12. Some of the earliest work on tutoring systems was done by Pat Suppes and colleagues at Stanford: Patrick Suppes and Mona Morningstar, "Computer-assisted instruction," *Science* 166 (1969): 343–50.
13. Michael Yudelson, Kenneth Koedinger, and Geoffrey Gordon, "Individualized Bayesian knowledge tracing models," in *Artificial Intelligence in Education: 16th International Conference*, ed. H. Chad Lane et al. (Springer, 2013).
14. For an example of machine learning on encrypted data, see, for example, Reza Shokri and Vitaly Shmatikov, "Privacy-preserving deep learning," in *Proceedings of the 22nd ACM SIGSAC Conference on Computer and Communications Security* (ACM, 2015).

15. A retrospective on the first smart home, based on a lecture by its inventor, James Sutherland: James E. Tomayko, "Electronic Computer for Home Operation (ECHO): The first home computer," *IEEE Annals of the History of Computing* 16 (1994): 59–61.

16. Summary of a smart-home project based on machine learning and automated decisions: Diane Cook et al., "MavHome: An agent-based smart home," in *Proceedings of the 1st IEEE International Conference on Pervasive Computing and Communications* (IEEE, 2003).

17. For the beginnings of an analysis of user experiences in smart homes, see Scott Davidoff et al., "Principles of smart home control," in *Ubicomp 2006: Ubiquitous Computing*, ed. Paul Dourish and Adrian Friday (Springer, 2006).

18. Commercial announcement of AI-based smart homes: "The Wolff Company unveils revolutionary smart home technology at new Annadel Apartments in Santa Rosa, California," *Business Insider*, March 12, 2018.

19. Article on robot chefs as commercial products: Eustacia Huen, "The world's first home robotic chef can cook over 100 meals," *Forbes*, October 31, 2016.

20. Report from my Berkeley colleagues on deep RL for robotic motor control: Sergey Levine et al., "End-to-end training of deep visuomotor policies," *Journal of Machine Learning Research* 17 (2016): 1–40.

21. On the possibilities for automating the work of hundreds of thousands of warehouse workers: Tom Simonite, "Grasping robots compete to rule Amazon's warehouses," *Wired*, July 26, 2017.

22. I'm assuming a generous one laptop-CPU minute per page, or about 10^{11} operations. A third-generation tensor processing unit from Google runs at about 10^{17} operations per second, meaning that it can read a million pages per second, or about five hours for eighty million two-hundred-page books.

23. A 2003 study on the global volume of information production by all channels: Peter Lyman and Hal Varian, "How much information?" sims.berkeley.edu/research/projects /how-much-info-2003.

24. For details on the use of speech recognition by intelligence agencies, see Dan Froomkin, "How the NSA converts spoken words into searchable text," *The Intercept*, May 5, 2015.

25. Analysis of visual imagery from satellites is an enormous task: Mike Kim, "Mapping poverty from space with the World Bank," Medium.com, January 4, 2017. Kim estimates eight million people working 24/7, which converts to more than thirty million people working forty hours per week. I suspect this is an overestimate in practice, because the vast majority of the images would exhibit negligible change over the course of one day. On the other hand, the US intelligence community employs tens of thousands of people sitting in vast rooms staring at satellite images just to keep track of what's happening in small regions of interest; so one million people is probably about right for the whole world.

26. There is substantial progress towards a global observatory based on real-time satellite image data: David Jensen and Jillian Campbell, "Digital earth: Building, financing and governing a digital ecosystem for planetary data," white paper for the UN Science-Policy-Business Forum on the Environment, 2018.

27. Luke Muehlhauser has written extensively on AI predictions, and I am indebted to him for tracking down original sources for the quotations that follow. See Luke Muehlhauser, "What should we learn from past AI forecasts?" Open Philanthropy Project report, 2016.

28. A forecast of the arrival of human-level AI within twenty years: Herbert Simon, *The New Science of Management Decision* (Harper & Row, 1960).

29. A forecast of the arrival of human-level AI within a generation: Marvin Minsky, *Computation: Finite and Infinite Machines* (Prentice Hall, 1967).

30. John McCarthy's forecast of the arrival of human-level AI within "five to 500 years": Ian Shenker, "Brainy robots in our future, experts think," *Detroit Free Press*, September 30, 1977.

31. For a summary of surveys of AI researchers on their estimates for the arrival of human-level AI, see aiimpacts.org. An extended discussion of survey results on human-level

AI is given by Katja Grace et al., "When will AI exceed human performance? Evidence from AI experts," arXiv:1705.08807v3 (2018).

32. For a chart mapping raw computer power against brain power, see Ray Kurzweil, "The law of accelerating returns," Kurzweilai.net, March 7, 2001.

33. The Allen Institute's Project Aristo: allenai.org/aristo.

34. For an analysis of the knowledge required to perform well on fourth-grade tests of comprehension and common sense, see Peter Clark et al., "Automatic construction of inference-supporting knowledge bases," in *Proceedings of the Workshop on Automated Knowledge Base Construction* (2014), akbc.ws/2014.

35. The NELL project on machine reading is described by Tom Mitchell et al., "Never-ending learning," *Communications of the ACM* 61 (2018): 103–15.

36. The idea of bootstrapping inferences from text is due to Sergey Brin, "Extracting patterns and relations from the World Wide Web," in *The World Wide Web and Databases*, ed. Paolo Atzeni, Alberto Mendelzon, and Giansalvatore Mecca (Springer, 1998).

37. For a visualization of the black-hole collision detected by LIGO, see LIGO Lab Caltech, "Warped space and time around colliding black holes," February 11, 2016, youtube.com/watch?v=1agm33iEAuo.

38. The first publication describing observation of gravitational waves: Ben Abbott et al., "Observation of gravitational waves from a binary black hole merger," *Physical Review Letters* 116 (2016): 061102.

39. On babies as scientists: Alison Gopnik, Andrew Meltzoff, and Patricia Kuhl, *The Scientist in the Crib: Minds, Brains, and How Children Learn* (William Morrow, 1999).

40. A summary of several projects on automated scientific analysis of experimental data to discover laws: Patrick Langley et al., *Scientific Discovery: Computational Explorations of the Creative Processes* (MIT Press, 1987).

41. Some early work on machine learning guided by prior knowledge: Stuart Russell, *The Use of Knowledge in Analogy and Induction* (Pitman, 1989).

42. Goodman's philosophical analysis of induction remains a source of inspiration: Nelson Goodman, *Fact, Fiction, and Forecast* (University of London Press, 1954).

43. A veteran AI researcher complains about mysticism in the philosophy of science: Herbert Simon, "Explaining the ineffable: AI on the topics of intuition, insight and inspiration," in *Proceedings of the 14th International Conference on Artificial Intelligence*, ed. Chris Mellish (Morgan Kaufmann, 1995).

44. A survey of inductive logic programming by two originators of the field: Stephen Muggleton and Luc de Raedt, "Inductive logic programming: Theory and methods," *Journal of Logic Programming* 19–20 (1994): 629–79.

45. For an early mention of the importance of encapsulating complex operations as new primitive actions, see Alfred North Whitehead, *An Introduction to Mathematics* (Henry Holt, 1911).

46. Work demonstrating that a simulated robot can learn entirely by itself to stand up: John Schulman et al., "High-dimensional continuous control using generalized advantage estimation," arXiv:1506.02438 (2015). A video demonstration is available at youtube.com/watch?v=SHLuf2ZBQSw.

47. A description of a reinforcement learning system that learns to play a capture-the-flag video game: Max Jaderberg et al., "Human-level performance in first-person multiplayer games with population-based deep reinforcement learning," arXiv:1807.01281 (2018).

48. A view of AI progress over the next few years: Peter Stone et al., "Artificial intelligence and life in 2030," *One Hundred Year Study on Artificial Intelligence*, report of the 2015 Study Panel, 2016.

49. The media-fueled argument between Elon Musk and Mark Zuckerberg: Peter Holley, "Billionaire burn: Musk says Zuckerberg's understanding of AI threat 'is limited,'" *The Washington Post*, July 25, 2017.

50. On the value of search engines to individual users: Erik Brynjolfsson, Felix Eggers, and Avinash Gannamaneni, "Using massive online choice experiments to measure changes in well-being," working paper no. 24514, National Bureau of Economic Research, 2018.

51. Penicillin was discovered several times and its curative powers were described in medical publications, but no one seems to have noticed. See en.wikipedia.org/wiki/History_of_penicillin.

52. For a discussion of some of the more esoteric risks from omniscient, clairvoyant AI systems, see David Auerbach, "The most terrifying thought experiment of all time," *Slate*, July 17, 2014.

53. An analysis of some potential pitfalls in thinking about advanced AI: Kevin Kelly, "The myth of a superhuman AI," *Wired*, April 25, 2017.

54. Machines may share *some* aspects of cognitive structure with humans, particularly those aspects dealing with perception and manipulation of the physical world and the conceptual structures involved in natural language understanding. Their deliberative processes are likely to be quite different because of the enormous disparities in hardware.

55. According to 2016 survey data, the eighty-eighth percentile corresponds to $100,000 per year: American Community Survey, US Census Bureau, www.census.gov/programs-surveys/acs. For the same year, global per capita GDP was $10,133: National Accounts Main Aggregates Database, UN Statistics Division, unstats.un.org/unsd/snaama.

56. If the GDP growth phases in over ten years or twenty years, it's worth $9,400 trillion or $6,800 trillion, respectively—still nothing to sneeze at. On an interesting historical note, I. J. Good, who popularized the notion of an intelligence explosion (page 142), estimated the value of human-level AI to be at least "one megaKeynes," referring to the fabled economist John Maynard Keynes. The value of Keynes's contributions was estimated in 1963 as £100 billion, so a megaKeynes comes out to around $2,200,000 trillion in 2016 dollars. Good pinned the value of AI primarily on its potential to ensure that the human race survives indefinitely. Later, he came to wonder whether he should have added a minus sign.

57. The EU announced plans for $24 billion in research and development spending for the period 2019–20. See European Commission, "Artificial intelligence: Commission outlines a European approach to boost investment and set ethical guidelines," press release, April 25, 2018. China's long-term investment plan for AI, announced in 2017, envisages a core AI industry generating $150 billion annually by 2030. See, for example, Paul Mozur, "Beijing wants A.I. to be made in China by 2030," *The New York Times*, July 20, 2017.

58. See, for example, Rio Tinto's Mine of the Future program at riotinto.com/australia/pilbara/mine-of-the-future-9603.aspx.

59. A retrospective analysis of economic growth: Jan Luiten van Zanden et al., eds., *How Was Life? Global Well-Being since 1820* (OECD Publishing, 2014).

60. The desire for relative advantage over others, rather than an absolute quality of life, is a *positional good*; see Chapter 9.

CHAPTER 4

1. Wikipedia's article on the Stasi has several useful references on its workforce and its overall impact on East German life.

2. For details on Stasi files, see Cullen Murphy, *God's Jury: The Inquisition and the Making of the Modern World* (Houghton Mifflin Harcourt, 2012).

3. For a thorough analysis of AI surveillance systems, see Jay Stanley, *The Dawn of Robot Surveillance* (American Civil Liberties Union, 2019).

4. Recent books on surveillance and control include Shoshana Zuboff, *The Age of Surveillance Capitalism: The Fight for a Human Future at the New Frontier of Power* (PublicAffairs, 2019) and Roger McNamee, *Zucked: Waking Up to the Facebook Catastrophe* (Penguin Press, 2019).

5. News article on a blackmail bot: Avivah Litan, "Meet Delilah—the first insider threat Trojan," Gartner Blog Network, July 14, 2016.

6. For a low-tech version of human susceptibility to misinformation, in which an unsuspecting individual becomes convinced that the world is being destroyed by meteor

strikes, see *Derren Brown: Apocalypse*, "Part One," directed by Simon Dinsell, 2012, youtube.com/watch?v=o_CUrMJOxqs.

7. An economic analysis of reputation systems and their corruption is given by Steven Tadelis, "Reputation and feedback systems in online platform markets," *Annual Review of Economics* 8 (2016): 321–40.

8. Goodhart's law: "Any observed statistical regularity will tend to collapse once pressure is placed upon it for control purposes." For example, there may once have been a correlation between faculty quality and faculty salary, so the *US News & World Report* college rankings measure faculty quality by faculty salaries. This has contributed to a salary arms race that benefits faculty members but not the students who pay for those salaries. The arms race changes faculty salaries in a way that does not depend on faculty quality, so the correlation tends to disappear.

9. An article describing German efforts to police public discourse: Bernhard Rohleder, "Germany set out to delete hate speech online. Instead, it made things worse," *WorldPost*, February 20, 2018.

10. On the "infopocalypse": Aviv Ovadya, "What's worse than fake news? The distortion of reality itself," *WorldPost*, February 22, 2018.

11. On the corruption of online hotel reviews: Dina Mayzlin, Yaniv Dover, and Judith Chevalier, "Promotional reviews: An empirical investigation of online review manipulation," *American Economic Review* 104 (2014): 2421–55.

12. Statement of Germany at the Meeting of the Group of Governmental Experts, Convention on Certain Conventional Weapons, Geneva, April 10, 2018.

13. The *Slaughterbots* movie, funded by the Future of Life Institute, appeared in November 2017 and is available at youtube.com/watch?v=9CO6M2HsoIA.

14. For a report on one of the bigger *faux pas* in military public relations, see Dan Lamothe, "Pentagon agency wants drones to hunt in packs, like wolves," *The Washington Post*, January 23, 2015.

15. Announcement of a large-scale drone swarm experiment: US Department of Defense, "Department of Defense announces successful micro-drone demonstration," news release no. NR-008-17, January 9, 2017.

16. Examples of research centers studying the impact of technology on employment are the Work and Intelligent Tools and Systems group at Berkeley, the Future of Work and Workers project at the Center for Advanced Study in the Behavioral Sciences at Stanford, and the Future of Work Initiative at Carnegie Mellon University.

17. A pessimistic take on future technological unemployment: Martin Ford, *Rise of the Robots: Technology and the Threat of a Jobless Future* (Basic Books, 2015).

18. Calum Chace, *The Economic Singularity: Artificial Intelligence and the Death of Capitalism* (Three Cs, 2016).

19. For an excellent collection of essays, see Ajay Agrawal, Joshua Gans, and Avi Goldfarb, eds., *The Economics of Artificial Intelligence: An Agenda* (National Bureau of Economic Research, 2019).

20. The mathematical analysis behind this "inverted-U" employment curve is given by James Bessen, "Artificial intelligence and jobs: The role of demand" in *The Economics of Artificial Intelligence*, ed. Agrawal, Gans, and Goldfarb.

21. For a discussion of economic dislocation arising from automation, see Eduardo Porter, "Tech is splitting the US work force in two," *The New York Times*, February 4, 2019. The article cites the following report for this conclusion: David Autor and Anna Salomons, "Is automation labor-displacing? Productivity growth, employment, and the labor share," *Brookings Papers on Economic Activity* (2018).

22. For data on the growth of banking in the twentieth century, see Thomas Philippon, "The evolution of the US financial industry from 1860 to 2007: Theory and evidence," working paper, 2008.

23. The bible for jobs data and the growth and decline of occupations: US Bureau of Labor Statistics, *Occupational Outlook Handbook: 2018–2019 Edition* (Bernan Press, 2018).

24. A report on trucking automation: Lora Kolodny, "Amazon is hauling cargo in self-driving trucks developed by Embark," CNBC, January 30, 2019.

25. The progress of automation in legal analytics, describing the results of a contest: Jason Tashea, "AI software is more accurate, faster than attorneys when assessing NDAs," *ABA Journal*, February 26, 2018.

26. A commentary by a distinguished economist, with a title explicitly evoking Keynes's 1930 article: Lawrence Summers, "Economic possibilities for our children," *NBER Reporter* (2013).

27. The analogy between data science employment and a small lifeboat for a giant cruise ship comes from a discussion with Yong Ying-I, head of Singapore's Public Service Division. She conceded that it was correct on the global scale, but noted that "Singapore is small enough to fit in the lifeboat."

28. Support for UBI from a conservative viewpoint: Sam Bowman, "The ideal welfare system is a basic income," Adam Smith Institute, November 25, 2013.

29. Support for UBI from a progressive viewpoint: Jonathan Bartley, "The Greens endorse a universal basic income. Others need to follow," *The Guardian*, June 2, 2017.

30. Chace, in *The Economic Singularity*, calls the "paradise" version of UBI the *Star Trek economy*, noting that in the more recent series of *Star Trek* episodes, money has been abolished because technology has created essentially unlimited material goods and energy. He also points to the massive changes in economic and social organization that will be needed to make such a system successful.

31. The economist Richard Baldwin also predicts a future of personal services in his book *The Globotics Upheaval: Globalization, Robotics, and the Future of Work* (Oxford University Press, 2019).

32. The book that is viewed as having exposed the failure of "whole-word" literacy education and launched decades of struggle between the two main schools of thought on reading: Rudolf Flesch, *Why Johnny Can't Read: And What You Can Do about It* (Harper & Bros., 1955).

33. On educational methods that enable the recipient to adapt to the rapid rate of technological and economic change in the next few decades: Joseph Aoun, *Robot-Proof: Higher Education in the Age of Artificial Intelligence* (MIT Press, 2017).

34. A radio lecture in which Turing predicted that humans would be overtaken by machines: Alan Turing, "Can digital machines think?," May 15, 1951, radio broadcast, BBC Third Programme. Typescript available at turingarchive.org.

35. News article describing the "naturalization" of Sophia as a citizen of Saudi Arabia: Dave Gershgorn, "Inside the mechanical brain of the world's first robot citizen," *Quartz*, November 12, 2017.

36. On Yann LeCun's view of Sophia: Shona Ghosh, "Facebook's AI boss described Sophia the robot as 'complete b——t' and 'Wizard-of-Oz AI,'" *Business Insider*, January 6, 2018.

37. An EU proposal on legal rights for robots: Committee on Legal Affairs of the European Parliament, "Report with recommendations to the Commission on Civil Law Rules on Robotics (2015/2103(INL))," 2017.

38. The GDPR provision on a "right to an explanation" is not, in fact, new: it is very similar to Article 15(1) of the 1995 Data Protection Directive, which it supersedes.

39. Here are three recent papers providing insightful mathematical analyses of fairness: Moritz Hardt, Eric Price, and Nati Srebro, "Equality of opportunity in supervised learning," in *Advances in Neural Information Processing Systems 29*, ed. Daniel Lee et al. (2016); Matt Kusner et al., "Counterfactual fairness," in *Advances in Neural Information Processing Systems 30*, ed. Isabelle Guyon et al. (2017); Jon Kleinberg, Sendhil Mullainathan, and Manish Raghavan, "Inherent trade-offs in the fair determination of risk scores," in *8th Innovations in Theoretical Computer Science Conference*, ed. Christos Papadimitriou (Dagstuhl Publishing, 2017).

40. News article describing the consequences of software failure for air traffic control: Simon Calder, "Thousands stranded by flight cancellations after systems failure at Europe's air-traffic coordinator," *The Independent*, April 3, 2018.

CHAPTER 5

1. Lovelace wrote, "The Analytical Engine has no pretensions whatever to originate anything. It can do whatever we know how to order it to perform. It can follow analysis; but it has no power of anticipating any analytical relations or truths." This was one of the arguments against AI that was refuted by Alan Turing, "Computing machinery and intelligence," *Mind* 59 (1950): 433–60.

2. The earliest known article on existential risk from AI was by Richard Thornton, "The age of machinery," *Primitive Expounder* IV (1847): 281.

3. "The Book of the Machines" was based on an earlier article by Samuel Butler, "Darwin among the machines," *The Press* (Christchurch, New Zealand), June 13, 1863.

4. Another lecture in which Turing predicted the subjugation of humankind: Alan Turing, "Intelligent machinery, a heretical theory" (lecture given to the 51 Society, Manchester, 1951). Typescript available at turingarchive.org.

5. Wiener's prescient discussion of technological control over humanity and a plea to retain human autonomy: Norbert Wiener, *The Human Use of Human Beings* (Riverside Press, 1950).

6. The front-cover blurb from Wiener's 1950 book is remarkably similar to the motto of the Future of Life Institute, an organization dedicated to studying the existential risks that humanity faces: "Technology is giving life the potential to flourish like never before . . . or to self-destruct."

7. An updating of Wiener's views arising from his increased appreciation of the possibility of intelligent machines: Norbert Wiener, *God and Golem, Inc.: A Comment on Certain Points Where Cybernetics Impinges on Religion* (MIT Press, 1964).

8. Asimov's Three Laws of Robotics first appeared in Isaac Asimov, "Runaround," *Astounding Science Fiction*, March 1942. The laws are as follows:

   1. A robot may not injure a human being or, through inaction, allow a human being to come to harm.
   2. A robot must obey the orders given it by human beings except where such orders would conflict with the First Law.
   3. A robot must protect its own existence as long as such protection does not conflict with the First or Second Laws.

   It is important to understand that Asimov proposed these laws as a way to generate interesting story plots, not as a serious guide for future roboticists. Several of his stories, including "Runaround," illustrate the problematic consequences of taking the laws literally. From the standpoint of modern AI, the laws fail to acknowledge any element of probability and risk: the legality of robot actions that expose a human to some probability of harm—however infinitesimal—is therefore unclear.

9. The notion of instrumental goals is due to Stephen Omohundro, "The nature of self-improving artificial intelligence" (unpublished manuscript, 2008). See also Stephen Omohundro, "The basic AI drives," in *Artificial General Intelligence 2008: Proceedings of the First AGI Conference*, ed. Pei Wang, Ben Goertzel, and Stan Franklin (IOS Press, 2008).

10. The objective of Johnny Depp's character, Will Caster, seems to be to solve the problem of physical reincarnation so that he can be reunited with his wife, Evelyn. This just goes to show that the nature of the overarching objective doesn't matter—the instrumental goals are all the same.

11. The original source for the idea of an intelligence explosion: I. J. Good, "Speculations concerning the first ultraintelligent machine," in *Advances in Computers*, vol. 6, ed. Franz Alt and Morris Rubinoff (Academic Press, 1965).

12. An example of the impact of the intelligence explosion idea: Luke Muehlhauser, in *Facing the Intelligence Explosion* (intelligenceexplosion.com), writes, "Good's paragraph ran over me like a train."

13. Diminishing returns can be illustrated as follows: suppose that a 16 percent improvement in intelligence creates a machine capable of making an 8 percent improvement, which in turn creates a 4 percent improvement, and so on. This process reaches a limit at about 36 percent above the original level. For more discussion on these issues, see Eliezer Yudkowsky, "Intelligence explosion microeconomics," technical report 2013-1, Machine Intelligence Research Institute, 2013.

14. For a view of AI in which humans become irrelevant, see Hans Moravec, *Mind Children: The Future of Robot and Human Intelligence* (Harvard University Press, 1988). See also Hans Moravec, *Robot: Mere Machine to Transcendent Mind* (Oxford University Press, 2000).

CHAPTER 6

1. A serious publication provides a serious review of Bostrom's *Superintelligence: Paths, Dangers, Strategies*: "Clever cogs," *Economist*, August 9, 2014.

2. A discussion of myths and misunderstandings concerning the risks of AI: Scott Alexander, "AI researchers on AI risk," *Slate Star Codex* (blog), May 22, 2015.

3. The classic work on multiple dimensions of intelligence: Howard Gardner, *Frames of Mind: The Theory of Multiple Intelligences* (Basic Books, 1983).

4. On the implications of multiple dimensions of intelligence for the possibility of superhuman AI: Kevin Kelly, "The myth of a superhuman AI," *Wired*, April 25, 2017.

5. Evidence that chimpanzees have better short-term memory than humans: Sana Inoue and Tetsuro Matsuzawa, "Working memory of numerals in chimpanzees," *Current Biology* 17 (2007), R1004–5.

6. An important early work questioning the prospects for rule-based AI systems: Hubert Dreyfus, *What Computers Can't Do* (MIT Press, 1972).

7. The first in a series of books seeking physical explanations for consciousness and raising doubts about the ability of AI systems to achieve real intelligence: Roger Penrose, *The Emperor's New Mind: Concerning Computers, Minds, and the Laws of Physics* (Oxford University Press, 1989).

8. A revival of the critique of AI based on the incompleteness theorem: Luciano Floridi, "Should we be afraid of AI?" *Aeon*, May 9, 2016.

9. A revival of the critique of AI based on the Chinese room argument: John Searle, "What your computer can't know," *The New York Review of Books*, October 9, 2014.

10. A report from distinguished AI researchers claiming that superhuman AI is probably impossible: Peter Stone et al., "Artificial intelligence and life in 2030," One Hundred Year Study on Artificial Intelligence, report of the 2015 Study Panel, 2016.

11. News article based on Andrew Ng's dismissal of risks from AI: Chris Williams, "AI guru Ng: Fearing a rise of killer robots is like worrying about overpopulation on Mars," *Register*, March 19, 2015.

12. An example of the "experts know best" argument: Oren Etzioni, "It's time to intelligently discuss artificial intelligence," *Backchannel*, December 9, 2014.

13. News article claiming that real AI researchers dismiss talk of risks: Erik Sofge, "Bill Gates fears AI, but AI researchers know better," *Popular Science*, January 30, 2015.

14. Another claim that real AI researchers dismiss AI risks: David Kenny, "IBM's open letter to Congress on artificial intelligence," June 27, 2017, ibm.com/blogs/policy/kenny-artificial-intelligence-letter.

15. Report from the workshop that proposed voluntary restrictions on genetic engineering: Paul Berg et al., "Summary statement of the Asilomar Conference on Recombinant DNA Molecules," *Proceedings of the National Academy of Sciences* 72 (1975): 1981–84.

16. Policy statement arising from the invention of CRISPR-Cas9 for gene editing: Organizing Committee for the International Summit on Human Gene Editing, "On human gene editing: International Summit statement," December 3, 2015.

17. The latest policy statement from leading biologists: Eric Lander et al., "Adopt a moratorium on heritable genome editing," *Nature* 567 (2019): 165–68.

18. Etzioni's comment that one cannot mention risks if one does not also mention benefits appears alongside his analysis of survey data from AI researchers: Oren Etzioni, "No, the experts don't think superintelligent AI is a threat to humanity," *MIT Technology Review*, September 20, 2016. In his analysis he argues that anyone who expects super-human AI to take more than twenty-five years—which includes this author as well as Nick Bostrom—is not concerned about the risks of AI.

19. A news article with quotations from the Musk–Zuckerberg "debate": Alanna Petroff, "Elon Musk says Mark Zuckerberg's understanding of AI is 'limited,'" *CNN Money*, July 25, 2017.

20. In 2015 the Information Technology and Innovation Foundation organized a debate titled "Are super intelligent computers really a threat to humanity?" Robert Atkinson, director of the foundation, suggests that mentioning risks is likely to result in reduced funding for AI. Video available at itif.org/events/2015/06/30/are-super-intelligent -computers-really-threat-humanity; the relevant discussion begins at 41:30.

21. A claim that our culture of safety will solve the AI control problem without ever mentioning it: Steven Pinker, "Tech prophecy and the underappreciated causal power of ideas," in *Possible Minds: Twenty-Five Ways of Looking at AI*, ed. John Brockman (Penguin Press, 2019).

22. For an interesting analysis of Oracle AI, see Stuart Armstrong, Anders Sandberg, and Nick Bostrom, "Thinking inside the box: Controlling and using an Oracle AI," *Minds and Machines* 22 (2012): 299–324.

23. Views on why AI is not going to take away jobs: Kenny, "IBM's open letter."

24. An example of Kurzweil's positive views of merging human brains with AI: Ray Kurzweil, interview by Bob Pisani, June 5, 2015, Exponential Finance Summit, New York, NY.

25. Article quoting Elon Musk on neural lace: Tim Urban, "Neuralink and the brain's magical future," Wait But Why, April 20, 2017.

26. For the most recent developments in Berkeley's neural dust project, see David Piech et al., "StimDust: A 1.7 mm^3, implantable wireless precision neural stimulator with ul-trasonic power and communication," arXiv: 1807.07590 (2018).

27. Susan Schneider, in *Artificial You: AI and the Future of Your Mind* (Princeton University Press, 2019), points out the risks of ignorance in proposed technologies such as uploading and neural prostheses: that, absent any real understanding of whether elec-tronic devices can be conscious and given the continuing philosophical confusion over persistent personal identity, we may inadvertently end our own conscious existences or inflict suffering on conscious machines without realizing that they are conscious.

28. An interview with Yann LeCun on AI risks: Guia Marie Del Prado, "Here's what Face-book's artificial intelligence expert thinks about the future," *Business Insider*, Septem-ber 23, 2015.

29. A diagnosis of AI control problems arising from an excess of testosterone: Steven Pinker, "Thinking does not imply subjugating," in *What to Think About Machines That Think*, ed. John Brockman (Harper Perennial, 2015).

30. A seminal work on many philosophical topics, including the question of whether moral obligations may be perceived in the natural world: David Hume, *A Treatise of Human Nature* (John Noon, 1738).

31. An argument that a sufficiently intelligent machine cannot help but pursue human objectives: Rodney Brooks, "The seven deadly sins of AI predictions," *MIT Technology Review*, October 6, 2017.

32. Pinker, "Thinking does not imply subjugating."

33. For an optimistic view arguing that AI safety problems will necessarily be resolved in our favor: Steven Pinker, "Tech prophecy."

34. On the unsuspected alignment between "skeptics" and "believers" in AI risk: Alexan-der, "AI researchers on AI risk."

NOTES

CHAPTER 7

1. For a guide to detailed brain modeling, now slightly outdated, see Anders Sandberg and Nick Bostrom, "*Whole brain emulation: A roadmap*," technical report 2008-3, Future of Humanity Institute, Oxford University, 2008.

2. For an introduction to genetic programming from a leading exponent, see John Koza, *Genetic Programming: On the Programming of Computers by Means of Natural Selection* (MIT Press, 1992).

3. The parallel to Asimov's Three Laws of Robotics is entirely coincidental.

4. The same point is made by Eliezer Yudkowsky, "Coherent extrapolated volition," technical report, Singularity Institute, 2004. Yudkowsky argues that directly building in "Four Great Moral Principles That Are All We Need to Program into AIs" is a sure road to ruin for humanity. His notion of the "coherent extrapolated volition of humankind" has the same general flavor as the first principle; the idea is that a superintelligent AI system could work out what humans, collectively, really want.

5. You can certainly have preferences over whether a machine is helping you achieve your preferences or you are achieving them through your own efforts. For example, suppose you prefer outcome A to outcome B, all other things being equal. You are unable to achieve outcome A unaided, and yet you still prefer B to getting A with the machine's help. In that case the machine should decide not to help you—unless perhaps it can do so in a way that is completely undetectable by you. You may, of course, have preferences about undetectable help as well as detectable help.

6. The phrase "the greatest good of the greatest number" originates in the work of Francis Hutcheson, *An Inquiry into the Original of Our Ideas of Beauty and Virtue, In Two Treatises* (D. Midwinter et al., 1725). Some have ascribed the formulation to an earlier comment by Wilhelm Leibniz; see Joachim Hruschka, "The greatest happiness principle and other early German anticipations of utilitarian theory," *Utilitas* 3 (1991): 165–77.

7. One might propose that the machine should include terms for animals as well as humans in its own objective function. If these terms have weights that correspond to how much people care about animals, then the end result will be the same as if the machine cares about animals only through caring about humans who care about animals. Giving each living animal equal weight in the machine's objective function would certainly be catastrophic—for example, we are outnumbered fifty thousand to one by Antarctic krill and a billion trillion to one by bacteria.

8. The moral philosopher Toby Ord made the same point to me in his comments on an early draft of this book: "Interestingly, the same is true in the study of moral philosophy. Uncertainty about moral value of outcomes was almost completely neglected in moral philosophy until very recently. Despite the fact that it is our uncertainty of moral matters that leads people to ask others for moral advice and, indeed, to do research on moral philosophy at all!"

9. One excuse for not paying attention to uncertainty about preferences is that it is formally equivalent to ordinary uncertainty, in the following sense: being uncertain about what I like is the same as being certain that I like likable things while being uncertain about what things are likable. This is just a trick that appears to move the uncertainty into the world, by making "likability by me" a property of objects rather than a property of me. In game theory, this trick has been thoroughly institutionalized since the 1960s, following a series of papers by my late colleague and Nobel laureate John Harsanyi: "Games with incomplete information played by 'Bayesian' players, Parts I–III," *Management Science* 14 (1967, 1968): 159–82, 320–34, 486–502. In decision theory, the standard reference is the following: Richard Cyert and Morris de Groot, "Adaptive utility," in *Expected Utility Hypotheses and the Allais Paradox*, ed. Maurice Allais and Ole Hagen (D. Reidel, 1979).

10. AI researchers working in the area of preference elicitation are an obvious exception. See, for example, Craig Boutilier, "On the foundations of *expected* expected utility," in *Proceedings of the 18th International Joint Conference on Artificial Intelligence* (Morgan Kaufmann, 2003). Also Alan Fern et al., "A decision-theoretic model of assistance," *Journal of Artificial Intelligence Research* 50 (2014): 71–104.

11. A critique of beneficial AI based on a misinterpretation of a journalist's brief interview with the author in a magazine article: Adam Elkus, "How to be good: Why you can't teach human values to artificial intelligence," *Slate*, April 20, 2016.

12. The origin of trolley problems: Frank Sharp, "A study of the influence of custom on the moral judgment," *Bulletin of the University of Wisconsin* 236 (1908).

13. The "anti-natalist" movement believes it is morally wrong for humans to reproduce because to live is to suffer and because humans' impact on the Earth is profoundly negative. If you consider the existence of humanity to be a moral dilemma, then I suppose I do want machines to resolve this moral dilemma the right way.

14. Statement on China's AI policy by Fu Ying, vice chair of the Foreign Affairs Committee of the National People's Congress. In a letter to the 2018 World AI Conference in Shanghai, Chinese president Xi Jinping wrote, "Deepened international cooperation is required to cope with new issues in fields including law, security, employment, ethics and governance." I am indebted to Brian Tse for bringing these statements to my attention.

15. A very interesting paper on the non-naturalistic non-fallacy, showing how preferences can be inferred from the state of the world as arranged by humans: Rohin Shah et al., "The implicit preference information in an initial state," in *Proceedings of the 7th International Conference on Learning Representations* (2019), iclr.cc/Conferences/2019/Schedule.

16. Retrospective on Asilomar: Paul Berg, "Asilomar 1975: DNA modification secured," *Nature* 455 (2008): 290–91.

17. News article reporting Putin's speech on AI: "Putin: Leader in artificial intelligence will rule world," Associated Press, September 4, 2017.

CHAPTER 8

1. Fermat's Last Theorem asserts that the equation $a^n = b^n + c^n$ has no solutions with a, b, and c being whole numbers and n being a whole number larger than 2. In the margin of his copy of Diophantus's *Arithmetica*, Fermat wrote, "I have a truly marvellous proof of this proposition which this margin is too narrow to contain." True or not, this guaranteed that mathematicians pursued a proof with vigor in the subsequent centuries. We can easily check particular cases—for example, is 7^3 equal to $6^3 + 5^3$? (Almost, because 7^3 is 343 and $6^3 + 5^3$ is 341, but "almost" doesn't count.) There are, of course, infinitely many cases to check, and that's why we need mathematicians and not just computer programmers.

2. A paper from the Machine Intelligence Research Institute poses many related issues: Scott Garrabrant and Abram Demski, "Embedded agency," AI Alignment Forum, November 15, 2018.

3. The classic work on multiattribute utility theory: Ralph Keeney and Howard Raiffa, *Decisions with Multiple Objectives: Preferences and Value Tradeoffs* (Wiley, 1976).

4. Paper introducing the idea of inverse RL: Stuart Russell, "Learning agents for uncertain environments," in *Proceedings of the 11th Annual Conference on Computational Learning Theory* (ACM, 1998).

5. The original paper on structural estimation of Markov decision processes: Thomas Sargent, "Estimation of dynamic labor demand schedules under rational expectations," *Journal of Political Economy* 86 (1978): 1009–44.

6. The first algorithms for IRL: Andrew Ng and Stuart Russell, "Algorithms for inverse reinforcement learning," in *Proceedings of the 17th International Conference on Machine Learning*, ed. Pat Langley (Morgan Kaufmann, 2000).

7. Better algorithms for inverse RL: Pieter Abbeel and Andrew Ng, "Apprenticeship learning via inverse reinforcement learning," in *Proceedings of the 21st International Conference on Machine Learning*, ed. Russ Greiner and Dale Schuurmans (ACM Press, 2004).

8. Understanding inverse RL as Bayesian updating: Deepak Ramachandran and Eyal Amir, "Bayesian inverse reinforcement learning," in *Proceedings of the 20th International Joint Conference on Artificial Intelligence*, ed. Manuela Veloso (AAAI Press, 2007).

9. How to teach helicopters to fly and do aerobatic maneuvers: Adam Coates, Pieter Abbeel, and Andrew Ng, "Apprenticeship learning for helicopter control," *Communications of the ACM* 52 (2009): 97–105.

10. The original name proposed for an assistance game was a *cooperative inverse reinforcement learning* game, or CIRL game. See Dylan Hadfield-Menell et al., "Cooperative inverse reinforcement learning," in *Advances in Neural Information Processing Systems 29*, ed. Daniel Lee et al. (2016).

11. These numbers are chosen just to make the game interesting.

12. The equilibrium solution to the game can be found by a process called *iterated best response*: pick any strategy for Harriet; pick the best strategy for Robbie, given Harriet's strategy; pick the best strategy for Harriet, given Robbie's strategy; and so on. If this process reaches a fixed point, where neither strategy changes, then we have found a solution. The process unfolds as follows:

   1. Start with the greedy strategy for Harriet: make 2 paperclips if she prefers paperclips; make 1 of each if she is indifferent; make 2 staples if she prefers staples.
   2. There are three possibilities Robbie has to consider, given this strategy for Harriet:
      a. If Robbie sees Harriet make 2 paperclips, he infers that she prefers paperclips, so he now believes the value of a paperclip is uniformly distributed between 50¢ and $1.00, with an average of 75¢. In that case, his best plan is to make 90 paperclips with an expected value of $67.50 for Harriet.
      b. If Robbie sees Harriet make 1 of each, he infers that she values paperclips and staples at 50¢, so the best choice is to make 50 of each.
      c. If Robbie sees Harriet make 2 staples, then by the same argument as in 2(a), he should make 90 staples.
   3. Given this strategy for Robbie, Harriet's best strategy is now somewhat different from the greedy strategy in step 1: if Robbie is going to respond to her making 1 of each by making 50 of each, then she is better off making 1 of each not just if she is *exactly* indifferent but if she is *anywhere close* to indifferent. In fact, the optimal policy is now to make 1 of each if she values paperclips anywhere between about 44.6¢ and 55.4¢.
   4. Given this new strategy for Harriet, Robbie's strategy remains unchanged. For example, if she chooses 1 of each, he infers that the value of a paperclip is uniformly distributed between 44.6¢ and 55.4¢, with an average of 50¢, so the best choice is to make 50 of each. Because Robbie's strategy is the same as in step 2, Harriet's best response will be the same as in step 3, and we have found the equilibrium.

13. For a more complete analysis of the off-switch game, see Dylan Hadfield-Menell et al., "The off-switch game," in *Proceedings of the 26th International Joint Conference on Artificial Intelligence*, ed. Carles Sierra (IJCAI, 2017).

14. The proof of the general result is quite simple if you don't mind integral signs. Let $P(u)$ be Robbie's prior probability density over Harriet's utility for the proposed action a. Then the value of going ahead with a is

$$EU(a) = \int_{-\infty}^{\infty} P(u) \cdot u \, du = \int_{-\infty}^{0} P(u) \cdot u \, du + \int_{0}^{\infty} P(u) \cdot u \, du$$

(We will see shortly why the integral is split up in this way.) On the other hand, the value of action d, deferring to Harriet, is composed of two parts: if $u > 0$, then Harriet lets Robbie go ahead, so the value is u, but if $u < 0$, then Harriet switches Robbie off, so the value is 0:

$$EU(d) = \int_{-\infty}^{0} P(u) \cdot 0 \, du + \int_{0}^{\infty} P(u) \cdot u \, du$$

Comparing the expressions for $EU(a)$ and $EU(d)$, we see immediately that $EU(d) \geq EU(a)$ because the expression for $EU(d)$ has the negative-utility region zeroed out. The two choices have equal value only when the negative region has zero probability—that is, when Robbie is already certain that Harriet likes the proposed action. The theorem is a direct analog of the well-known theorem concerning the non-negative expected value of information.

15. Perhaps the next elaboration in line, for the one human–one robot case, is to consider a Harriet who does not yet know her own preferences regarding some aspect of the world, or whose preferences have not yet been formed.

16. To see how exactly Robbie converges to an incorrect belief, consider a model in which Harriet is slightly irrational, making errors with a probability that diminishes exponentially as the size of error increases. Robbie offers Harriet 4 paperclips in return for 1 staple; she refuses. According to Robbie's beliefs, this is irrational: even at 25¢ per paperclip and 75¢ per staple, she should accept 4 for 1. Therefore, she must have made a mistake—but this mistake is *much* more likely if her true value is 25¢ than if it is, say, 30¢, because the error costs her a lot more if her value for paperclips is 30¢. Now Robbie's probability distribution has 25¢ as the most likely value because it represents the smallest error on Harriet's part, with exponentially lower probabilities for values higher than 25¢. If he keeps trying the same experiment, the probability distribution becomes more and more concentrated close to 25¢. In the limit, Robbie becomes certain that Harriet's value for paperclips is 25¢.

17. Robbie could, for example, have a normal (Gaussian) distribution for his prior belief about the exchange rate, which stretches from $-\infty$ to $+\infty$.

18. For an example of the kind of mathematical analysis that may be needed, see Avrim Blum, Lisa Hellerstein, and Nick Littlestone, "Learning in the presence of finitely or infinitely many irrelevant attributes," *Journal of Computer and System Sciences* 50 (1995): 32–40. Also Lori Dalton, "Optimal Bayesian feature selection," in *Proceedings of the 2013 IEEE Global Conference on Signal and Information Processing*, ed. Charles Bouman, Robert Nowak, and Anna Scaglione (IEEE, 2013).

19. Here I am rephrasing slightly a question by Moshe Vardi at the Asilomar Conference on Beneficial AI, 2017.

20. Michael Wellman and Jon Doyle, "Preferential semantics for goals," in *Proceedings of the 9th National Conference on Artificial Intelligence* (AAAI Press, 1991). This paper draws on a much earlier proposal by Georg von Wright, "The logic of preference reconsidered," *Theory and Decision* 3 (1972): 140–67.

21. My late Berkeley colleague has the distinction of becoming an adjective. See Paul Grice, *Studies in the Way of Words* (Harvard University Press, 1989).

22. The original paper on direct stimulation of pleasure centers in the brain: James Olds and Peter Milner, "Positive reinforcement produced by electrical stimulation of septal area and other regions of rat brain," *Journal of Comparative and Physiological Psychology* 47 (1954): 419–27.

23. Letting rats push the button: James Olds, "Self-stimulation of the brain; its use to study local effects of hunger, sex, and drugs," *Science* 127 (1958): 315–24.

24. Letting humans push the button: Robert Heath, "Electrical self-stimulation of the brain in man," *American Journal of Psychiatry* 120 (1963): 571–77.

25. A first mathematical treatment of wireheading, showing how it occurs in reinforcement learning agents: Mark Ring and Laurent Orseau, "Delusion, survival, and intelligent agents," in *Artificial General Intelligence: 4th International Conference*, ed. Jürgen Schmidhuber, Kristinn Thórisson, and Moshe Looks (Springer, 2011). One possible solution to the wireheading problem: Tom Everitt and Marcus Hutter, "Avoiding wireheading with value reinforcement learning," arXiv:1605.03143 (2016).

26. How it might be possible for an intelligence explosion to occur safely: Benja Fallenstein and Nate Soares, "Vingean reflection: Reliable reasoning for self-improving agents," technical report 2015-2, Machine Intelligence Research Institute, 2015.

27. The difficulty agents face in reasoning about themselves and their successors: Benja Fallenstein and Nate Soares, "Problems of self-reference in self-improving space-time embedded intelligence," in *Artificial General Intelligence: 7th International Conference*, ed. Ben Goertzel, Laurent Orseau, and Javier Snaider (Springer, 2014).

28. Showing why an agent might pursue an objective different from its true objective if its computational abilities are limited: Jonathan Sorg, Satinder Singh, and Richard Lewis, "Internal rewards mitigate agent boundedness," in *Proceedings of the 27th International Conference on Machine Learning*, ed. Johannes Fürnkranz and Thorsten Joachims (2010), icml.cc/Conferences/2010/papers/icml2010proceedings.zip.

CHAPTER 9

1. Some have argued that biology and neuroscience are also directly relevant. See, for example, Gopal Sarma, Adam Safron, and Nick Hay, "Integrative biological simulation, neuropsychology, and AI safety," arxiv.org/abs/1811.03493 (2018).

2. On the possibility of making computers liable for damages: Paulius Čerka, Jurgita Grigienė, and Gintarė Sirbikytė, "Liability for damages caused by artificial intelligence," *Computer Law and Security Review* 31 (2015): 376–89.

3. For an excellent machine-oriented introduction to standard ethical theories and their implications for designing AI systems, see Wendell Wallach and Colin Allen, *Moral Machines: Teaching Robots Right from Wrong* (Oxford University Press, 2008).

4. The sourcebook for utilitarian thought: Jeremy Bentham, *An Introduction to the Principles of Morals and Legislation* (T. Payne & Son, 1789).

5. Mill's elaboration of his tutor Bentham's ideas was extraordinarily influential on liberal thought: John Stuart Mill, *Utilitarianism* (Parker, Son & Bourn, 1863).

6. The paper introducing preference utilitarianism and preference autonomy: John Harsanyi, "Morality and the theory of rational behavior," *Social Research* 44 (1977): 623–56.

7. An argument for social aggregation via weighted sums of utilities when deciding on behalf of multiple individuals: John Harsanyi, "Cardinal welfare, individualistic ethics, and interpersonal comparisons of utility," *Journal of Political Economy* 63 (1955): 309–21.

8. A generalization of Harsanyi's social aggregation theorem to the case of unequal prior beliefs: Andrew Critch, Nishant Desai, and Stuart Russell, "Negotiable reinforcement learning for Pareto optimal sequential decision-making," in *Advances in Neural Information Processing Systems 31*, ed. Samy Bengio et al. (2018).

9. The sourcebook for ideal utilitarianism: G. E. Moore, *Ethics* (Williams & Norgate, 1912).

10. News article citing Stuart Armstrong's colorful example of misguided utility maximization: Chris Matyszczyk, "Professor warns robots could keep us in coffins on heroin drips," CNET, June 29, 2015.

11. Popper's theory of negative utilitarianism (so named later by Smart): Karl Popper, *The Open Society and Its Enemies* (Routledge, 1945).

12. A refutation of negative utilitarianism: R. Ninian Smart, "Negative utilitarianism," *Mind* 67 (1958): 542–43.

13. For a typical argument for risks arising from "end human suffering" commands, see "Why do we think AI will destroy us?," Reddit, reddit.com/r/Futurology/comments/38fp6o/why_do_we_think_ai_will_destroy_us.

14. A good source for self-deluding incentives in AI: Ring and Orseau, "Delusion, survival, and intelligent agents."

15. On the impossibility of interpersonal comparisons of utility: W. Stanley Jevons, *The Theory of Political Economy* (Macmillan, 1871).

16. The utility monster makes its appearance in Robert Nozick, *Anarchy, State, and Utopia* (Basic Books, 1974).

17. For example, we can fix immediate death to have a utility of 0 and a maximally happy life to have a utility of 1. See John Isbell, "Absolute games," in *Contributions to the Theory of Games*, vol. 4, ed. Albert Tucker and R. Duncan Luce (Princeton University Press, 1959).

18. The oversimplified nature of Thanos's population-halving policy is discussed by Tim Harford, "Thanos shows us how not to be an economist," *Financial Times*, April 20, 2019. Even before the film debuted, defenders of Thanos began to congregate on the subreddit r/thanosdidnothingwrong/. In keeping with the subreddit's motto, 350,000 of the 700,000 members were later purged.

19. On utilities for populations of different sizes: Henry Sidgwick, *The Methods of Ethics* (Macmillan, 1874).

20. The Repugnant Conclusion and other knotty problems of utilitarian thinking: Derek Parfit, *Reasons and Persons* (Oxford University Press, 1984).

21. For a concise summary of axiomatic approaches to population ethics, see Peter Eckersley, "Impossibility and uncertainty theorems in AI value alignment," in *Proceedings of the AAAI Workshop on Artificial Intelligence Safety*, ed. Huáscar Espinoza et al. (2019).

22. Calculating the long-term carrying capacity of the Earth: Daniel O'Neill et al., "A good life for all within planetary boundaries," *Nature Sustainability* 1 (2018): 88–95.

23. For an application of moral uncertainty to population ethics, see Hilary Greaves and Toby Ord, "Moral uncertainty about population axiology," *Journal of Ethics and Social Philosophy* 12 (2017): 135–67. A more comprehensive analysis is provided by Will MacAskill, Krister Bykvist, and Toby Ord, *Moral Uncertainty* (Oxford University Press, forthcoming).

24. Quotation showing that Smith was not so obsessed with selfishness as is commonly imagined: Adam Smith, *The Theory of Moral Sentiments* (Andrew Millar; Alexander Kincaid and J. Bell, 1759).

25. For an introduction to the economics of altruism, see Serge-Christophe Kolm and Jean Ythier, eds., *Handbook of the Economics of Giving, Altruism and Reciprocity*, 2 vols. (North-Holland, 2006).

26. On charity as selfish: James Andreoni, "Impure altruism and donations to public goods: A theory of warm-glow giving," *Economic Journal* 100 (1990): 464–77.

27. For those who like equations: let Alice's intrinsic well-being be measured by w_A and Bob's by w_B. Then the utilities for Alice and Bob are defined as follows:

$$U_A = w_A + C_{AB} \, w_B$$
$$U_B = w_B + C_{BA} \, w_A.$$

Some authors suggest that Alice cares about Bob's overall utility U_B rather than just his intrinsic well-being w_B, but this leads to a kind of circularity in that Alice's utility depends on Bob's utility, which depends on Alice's utility; sometimes stable solutions can be found but the underlying model can be questioned. See, for example, Hajime Hori, "Nonpaternalistic altruism and functional interdependence of social preferences," *Social Choice and Welfare* 32 (2009): 59–77.

28. Models in which each individual's utility is a linear combination of everyone's well-being are just one possibility. Much more general models are possible—for example, models in which some individuals prefer to avoid severe inequalities in the distribution of well-being, even at the expense of reducing the total, while other individuals would really prefer that no one have preferences about inequality at all. Thus, the overall approach I am proposing accommodates multiple moral theories held by individuals; at the same time, it doesn't insist that any one of those moral theories is correct or should have much sway over outcomes for those who hold a different theory. I am indebted to Toby Ord for pointing out this feature of the approach.

29. Arguments of this type have been made against policies designed to ensure equality of outcome, notably by the American legal philosopher Ronald Dworkin. See, for example, Ronald Dworkin, "What is equality? Part 1: Equality of welfare," *Philosophy and Public Affairs* 10 (1981): 185–246. I am indebted to Iason Gabriel for this reference.

30. Malice in the form of revenge-based punishment for transgressions is certainly a common tendency. Although it plays a social role in keeping members of a community in line, it can be replaced by an equally effective policy driven by deterrence and prevention—that is, weighing the intrinsic harm done when punishing the transgressor against the benefits to the larger society.

31. Let E_{AB} and P_{AB} be Alice's coefficients of envy and pride respectively, and assume that they apply to the difference in well-being. Then a (somewhat oversimplified) formula for Alice's utility could be the following:

$$U_A = w_A + C_{AB} \, w_B - E_{AB} \, (w_B - w_A) + P_{AB} \, (w_A - w_B)$$
$$= (1 + E_{AB} + P_{AB}) \, w_A + (C_{AB} - E_{AB} - P_{AB}) \, w_B.$$

Thus, if Alice has positive pride and envy coefficients, they act on Bob's welfare exactly like sadism and malice coefficients: Alice is happier if Bob's welfare is lowered, all other things being equal. In reality, pride and envy typically apply not to differences in well-being but to differences in visible aspects thereof, such as status and possessions. Bob's hard toil in acquiring his possessions (which lowers his overall well-being) may not be visible to Alice. This can lead to the self-defeating behaviors that go under the heading of "keeping up with the Joneses."

32. On the sociology of conspicuous consumption: Thorstein Veblen, *The Theory of the Leisure Class: An Economic Study of Institutions* (Macmillan, 1899).

33. Fred Hirsch, *The Social Limits to Growth* (Routledge & Kegan Paul, 1977).

34. I am indebted to Ziyad Marar for pointing me to social identity theory and its importance in understanding human motivation and behavior. See, for example, Dominic Abrams and Michael Hogg, eds., *Social Identity Theory: Constructive and Critical Advances* (Springer, 1990). For a much briefer summary of the main ideas, see Ziyad Marar, "Social identity," in *This Idea Is Brilliant: Lost, Overlooked, and Underappreciated Scientific Concepts Everyone Should Know*, ed. John Brockman (Harper Perennial, 2018).

35. Here, I am not suggesting that we necessarily need a detailed understanding of the neural implementation of cognition; what is needed is a model at the "software" level of how preferences, both explicit and implicit, generate behavior. Such a model would need to incorporate what is known about the reward system.

36. Ralph Adolphs and David Anderson, *The Neuroscience of Emotion: A New Synthesis* (Princeton University Press, 2018).

37. See, for example, Rosalind Picard, *Affective Computing*, 2nd ed. (MIT Press, 1998).

38. Waxing lyrical on the delights of the durian: Alfred Russel Wallace, *The Malay Archipelago: The Land of the Orang-Utan, and the Bird of Paradise* (Macmillan, 1869).

39. A less rosy view of the durian: Alan Davidson, *The Oxford Companion to Food* (Oxford University Press, 1999). Buildings have been evacuated and planes turned around in mid-flight because of the durian's overpowering odor.

40. I discovered after writing this chapter that the durian was used for exactly the same philosophical purpose by Laurie Paul, *Transformative Experience* (Oxford University Press, 2014). Paul suggests that uncertainty about one's own preferences presents fatal problems for decision theory, a view contradicted by Richard Pettigrew, "Transformative experience and decision theory," *Philosophy and Phenomenological Research* 91 (2015): 766–74. Neither author refers to the early work of Harsanyi, "Games with incomplete information, Parts I–III," or Cyert and de Groot, "Adaptive utility."

41. An initial paper on helping humans who don't know their own preferences and are learning about them: Lawrence Chan et al., "The assistive multi-armed bandit," in *Proceedings of the 14th ACM/IEEE International Conference on Human–Robot Interaction (HRI)*, ed. David Sirkin et al. (IEEE, 2019).

42. Eliezer Yudkowsky, in *Coherent Extrapolated Volition* (Singularity Institute, 2004), lumps all these aspects, as well as plain inconsistency, under the heading of *muddle*—a term that has not, unfortunately, caught on.

43. On the two selves who evaluate experiences: Daniel Kahneman, *Thinking, Fast and Slow* (Farrar, Straus & Giroux, 2011).

44. Edgeworth's hedonimeter, an imaginary device for measuring happiness moment to moment: Francis Edgeworth, *Mathematical Psychics: An Essay on the Application of Mathematics to the Moral Sciences* (Kegan Paul, 1881).

45. A standard text on sequential decisions under uncertainty: Martin Puterman, *Markov Decision Processes: Discrete Stochastic Dynamic Programming* (Wiley, 1994).

46. On axiomatic assumptions that justify additive representations of utility over time: Tjalling Koopmans, "Representation of preference orderings over time," in *Decision and Organization*, ed. C. Bartlett McGuire, Roy Radner, and Kenneth Arrow (North-Holland, 1972).

47. The 2019 humans (who might, in 2099, be long dead or might just be the earlier selves of 2099 humans) might wish to build the machines in a way that respects the 2019 preferences of the 2019 humans rather than pandering to the undoubtedly shallow and ill-considered preferences of humans in 2099. This would be like drawing up a constitution that disallows any amendments. If the 2099 humans, after suitable deliberation, decide they wish to override the preferences built in by the 2019 humans, it seems reasonable that they should be able to do so. After all, it is they and their descendants who have to live with the consequences.

48. I am indebted to Wendell Wallach for this observation.
49. An early paper dealing with changes in preferences over time: John Harsanyi, "Welfare economics of variable tastes," *Review of Economic Studies* 21 (1953): 204–13. A more recent (and somewhat technical) survey is provided by Franz Dietrich and Christian List, "Where do preferences come from?," *International Journal of Game Theory* 42 (2013): 613–37. See also Laurie Paul, *Transformative Experience* (Oxford University Press, 2014), and Richard Pettigrew, "Choosing for Changing Selves," philpapers.org /archive/PETCFC.pdf.
50. For a rational analysis of irrationality, see Jon Elster, *Ulysses and the Sirens: Studies in Rationality and Irrationality* (Cambridge University Press, 1979).
51. For promising ideas on cognitive prostheses for humans, see Falk Lieder, "Beyond bounded rationality: Reverse-engineering and enhancing human intelligence" (PhD thesis, University of California, Berkeley, 2018).

CHAPTER 10

1. On the application of assistance games to driving: Dorsa Sadigh et al., "Planning for cars that coordinate with people," *Autonomous Robots* 42 (2018): 1405–26.
2. Apple is, curiously, absent from this list. It does have an AI research group and is ramping up rapidly. Its traditional culture of secrecy means that its impact in the marketplace of ideas is quite limited so far.
3. Max Tegmark, interview, *Do You Trust This Computer?*, directed by Chris Paine, written by Mark Monroe (2018).
4. On estimating the impact of cybercrime: "Cybercrime cost $600 billion and targets banks first," *Security Magazine*, February 21, 2018.

APPENDIX A

1. The basic plan for chess programs of the next sixty years: Claude Shannon, "Programming a computer for playing chess," *Philosophical Magazine*, 7th ser., 41 (1950): 256–75. Shannon's proposal drew on a centuries-long tradition of evaluating chess positions by adding up piece values; see, for example, Pietro Carrera, *Il gioco degli scacchi* (Giovanni de Rossi, 1617).
2. A report describing Samuel's heroic research on an early reinforcement learning algorithm for checkers: Arthur Samuel, "Some studies in machine learning using the game of checkers," *IBM Journal of Research and Development* 3 (1959): 210–29.
3. The concept of rational metareasoning and its application to search and game playing emerged from the thesis research of my student Eric Wefald, who died tragically in a car accident before he could write up his work; the following appeared posthumously: Stuart Russell and Eric Wefald, *Do the Right Thing: Studies in Limited Rationality* (MIT Press, 1991). See also Eric Horvitz, "Rational metareasoning and compilation for optimizing decisions under bounded resources," in *Computational Intelligence, II: Proceedings of the International Symposium*, ed. Francesco Gardin and Giancarlo Mauri (North-Holland, 1990); and Stuart Russell and Eric Wefald, "On optimal game-tree search using rational meta-reasoning," in *Proceedings of the 11th International Joint Conference on Artificial Intelligence*, ed. Natesa Sridharan (Morgan Kaufmann, 1989).
4. Perhaps the first paper showing how hierarchical organization reduces the combinatorial complexity of planning: Herbert Simon, "The architecture of complexity," *Proceedings of the American Philosophical Society* 106 (1962): 467–82.
5. The canonical reference for hierarchical planning is Earl Sacerdoti, "Planning in a hierarchy of abstraction spaces," *Artificial Intelligence* 5 (1974): 115–35. See also Austin Tate, "Generating project networks," in *Proceedings of the 5th International Joint Conference on Artificial Intelligence*, ed. Raj Reddy (Morgan Kaufmann, 1977).
6. A formal definition of what high-level actions do: Bhaskara Marthi, Stuart Russell, and Jason Wolfe, "Angelic semantics for high-level actions," in *Proceedings of the 17th International Conference on Automated Planning and Scheduling*, ed. Mark Boddy, Maria Fox, and Sylvie Thiébaux (AAAI Press, 2007).

APPENDIX B

1. This example is unlikely to be from Aristotle, but may have originated with Sextus Empiricus, who lived probably in the second or third century CE.
2. The first algorithm for theorem-proving in first-order logic worked by reducing first-order sentences to (very large numbers of) propositional sentences: Martin Davis and Hilary Putnam, "A computing procedure for quantification theory," *Journal of the ACM* 7 (1960): 201–15.
3. An improved algorithm for propositional inference: Martin Davis, George Logemann, and Donald Loveland, "A machine program for theorem-proving," *Communications of the ACM* 5 (1962): 394–97.
4. The satisfiability problem—deciding whether a collection of sentences is true in *some* world—is NP-complete. The reasoning problem—deciding whether a sentence follows from the known sentences—is co-NP-complete, a class that is thought to be harder than NP-complete problems.
5. There are two exceptions to this rule: no repetition (a stone may not be played that returns the board to a situation that existed previously) and no suicide (a stone may not be placed such that it would immediately be captured—for example, if it is already surrounded).
6. The work that introduced first-order logic as we understand it today (*Begriffsschrift* means "concept writing"): Gottlob Frege, *Begriffsschrift, eine der arithmetischen nachgebildete Formelsprache des reinen Denkens* (Halle, 1879). Frege's notation for first-order logic was so bizarre and unwieldy that it was soon replaced by the notation introduced by Giuseppe Peano, which remains in common use today.
7. A summary of Japan's bid for supremacy through knowledge-based systems: Edward Feigenbaum and Pamela McCorduck, *The Fifth Generation: Artificial Intelligence and Japan's Computer Challenge to the World* (Addison-Wesley, 1983).
8. The US efforts included the Strategic Computing Initiative and the formation of the Microelectronics and Computer Technology Corporation (MCC). See Alex Roland and Philip Shiman, *Strategic Computing: DARPA and the Quest for Machine Intelligence, 1983–1993* (MIT Press, 2002).
9. A history of Britain's response to the re-emergence of AI in the 1980s: Brian Oakley and Kenneth Owen, *Alvey: Britain's Strategic Computing Initiative* (MIT Press, 1990).
10. The origin of the term GOFAI: John Haugeland, *Artificial Intelligence: The Very Idea* (MIT Press, 1985).
11. Interview with Demis Hassabis on the future of AI and deep learning: Nick Heath, "Google DeepMind founder Demis Hassabis: Three truths about AI," *TechRepublic,* September 24, 2018.

APPENDIX C

1. Pearl's work was recognized by the Turing Award in 2011.
2. Bayes nets in more detail: Every node in the network is annotated with the probability of each possible value, given each possible combination of values for the node's *parents* (that is, those nodes that point to it). For example, the probability that $Doubles_{12}$ has value *true* is 1.0 when D_1 and D_2 have the same value, and 0.0 otherwise. A possible world is an assignment of values to all the nodes. The probability of such a world is the product of the appropriate probabilities from each of the nodes.
3. A compendium of applications of Bayes nets: Olivier Pourret, Patrick Naïm, and Bruce Marcot, eds., *Bayesian Networks: A Practical Guide to Applications* (Wiley, 2008).
4. The basic paper on probabilistic programming: Daphne Koller, David McAllester, and Avi Pfeffer, "Effective Bayesian inference for stochastic programs," in *Proceedings of the 14th National Conference on Artificial Intelligence* (AAAI Press, 1997). For many additional references, see probabilistic-programming.org.
5. Using probabilistic programs to model human concept learning: Brenden Lake, Ruslan Salakhutdinov, and Joshua Tenenbaum, "Human-level concept learning through probabilistic program induction," *Science* 350 (2015): 1332–38.

6. For a detailed description of the seismic monitoring application and associated probability model, see Nimar Arora, Stuart Russell, and Erik Sudderth, "NET-VISA: Network processing vertically integrated seismic analysis," *Bulletin of the Seismological Society of America* 103 (2013): 709–29.

7. News article describing one of the first serious self-driving car crashes: Ryan Randazzo, "Who was at fault in self-driving Uber crash? Accounts in Tempe police report disagree," *Republic* (azcentral.com), March 29, 2017.

APPENDIX D

1. The foundational discussion of inductive learning: David Hume, *Philosophical Essays Concerning Human Understanding* (A. Millar, 1748).

2. Leslie Valiant, "A theory of the learnable," *Communications of the ACM* 27 (1984): 1134–42. See also Vladimir Vapnik, *Statistical Learning Theory* (Wiley, 1998). Valiant's approach concentrated on computational complexity, Vapnik's on statistical analysis of the learning capacity of various classes of hypotheses, but both shared a common theoretical core connecting data and predictive accuracy.

3. For example, to learn the difference between the "situational superko" and "natural situational superko" rules, the learning algorithm would have to try repeating a board position that it had created previously by a pass rather than by playing a stone. The results would be different in different countries.

4. For a description of the ImageNet competition, see Olga Russakovsky et al., "ImageNet large scale visual recognition challenge," *International Journal of Computer Vision* 115 (2015): 211–52.

5. The first demonstration of deep networks for vision: Alex Krizhevsky, Ilya Sutskever, and Geoffrey Hinton, "ImageNet classification with deep convolutional neural networks," in *Advances in Neural Information Processing Systems 25*, ed. Fernando Pereira et al. (2012).

6. The difficulty of distinguishing over one hundred breeds of dogs: Andrej Karpathy, "What I learned from competing against a ConvNet on ImageNet," *Andrej Karpathy Blog*, September 2, 2014.

7. Blog post on inceptionism research at Google: Alexander Mordvintsev, Christopher Olah, and Mike Tyka, "Inceptionism: Going deeper into neural networks," *Google AI Blog*, June 17, 2015. The idea seems to have originated with J. P. Lewis, "Creation by refinement: A creativity paradigm for gradient descent learning networks," in *Proceedings of the IEEE International Conference on Neural Networks* (IEEE, 1988).

8. News article on Geoff Hinton having second thoughts about deep networks: Steve LeVine, "Artificial intelligence pioneer says we need to start over," *Axios*, September 15, 2017.

9. A catalog of shortcomings of deep learning: Gary Marcus, "Deep learning: A critical appraisal," arXiv:1801.00631 (2018).

10. A popular textbook on deep learning, with a frank assessment of its weaknesses: François Chollet, *Deep Learning with Python* (Manning Publications, 2017).

11. An explanation of explanation-based learning: Thomas Dietterich, "Learning at the knowledge level," *Machine Learning* 1 (1986): 287–315.

12. A superficially quite different explanation of explanation-based learning: John Laird, Paul Rosenbloom, and Allen Newell, "Chunking in Soar: The anatomy of a general learning mechanism," *Machine Learning* 1 (1986): 11–46.

Image Credits

Index